Android App-Hook and Plug-In Technology

Android App-Hook and Plug-In Technology

Jianqiang Bao

CRC Press
Taylor & Francis Group
Boca Raton London New York

CRC Press is an imprint of the
Taylor & Francis Group, an **informa** business

HZ BOOKS
华章科技

CRC Press
Taylor & Francis Group
6000 Broken Sound Parkway NW, Suite 300
Boca Raton, FL 33487-2742

CRC Press is an imprint of Taylor & Francis Group, an Informa business

International Standard Book Number-13: 978-0-367-20700-7 (Hardback)

Visit the Taylor & Francis Web site at
http://www.taylorandfrancis.com

and the CRC Press Web site at
http://www.crcpress.com

Contents

Acknowledgments

I OWE A BIG THANKS to Yi Wu, the editor of this book in Chinese, for helping me contact CRC Press to publish this book in English.

I thank Manyun Guo, my wife, for accompanying and encouraging me during the period I spent writing.

Special thanks to my 21 friends from Android forums for helping me translate this Chinese book into English.

I thank Yong Zhang, Yugang Ren, Lody, Guangliang Lin, Jian Huang, and a lot of other friends. Without your endless support I wouldn't have been able to talk as deeply about this technique domain.

About the Author

 **Jianqiang Bao** is a senior Android app developer. For more than 15 years, he has developed enterprise solutions using Silverlight, ASP.NET, WP7, Android, and iOS. He has worked at HP, Microsoft, Tuniu and Qunar. He has a technique blog at http://www.cnblogs.com/jax; his GitHub is https://github.com/BaoBaoJianqiang.

About the Translators

M ANY PEOPLE HAVE PARTICIPATED in this book's translation from Chinese to English, the list is as follows:

Chapter	Translator	Reviewer
1	Hongwei Cao	Han Yan
2	Chunfei Shi, Xuelong Wang, Xiaohui Li	Fangxiang Deng
3	Wenpeng Li	Jianqiang Bao
4	Xizhi Pan	Jinyu Guo
5	Jian Feng	Guiming Zou
6	Xiaohui Li	Zelong Gong
7	Jinyu Guo	Tong Peng
8	Tianhong Han	Wenhan Xiao
	Guiming Zou	Sheng Li
	Xuelong Wang	Jian Feng
	Yupeng Wang	
	Sheng Li	
9	Alan Pan T	Xizhi Pan
	Siyang Long	
10	Shuaifeng Ma	Tong Peng
	Fangxiang Deng	Jian Feng
	Tong Peng	Fangxiang Deng
	Zhaoyun	Siyang Long
	Zelong Gong	Guiming Zou
	Hao Yang	Xizhi Pan
	Jinyu Guo	
11	Jinyu Guo	Siyang Long

Introduction

WELCOME TO THE FIRST edition of *Android App-Hook and Plug-In Technology*

WHAT THIS BOOK WILL TEACH YOU

This book will teach you everything you need to know to master Android plug-in techniques.

This book introduces the Android plug-in technique. An app can be downloaded as an apk file in a zip file from the remote server. We call this zip file a plug-in. This app can invoke a class in this plug-in. This means that we can update the content of the app without republishing it again.

Google Play has a strict app auditing strategy. It forbids any app from downloading another app to prevent it from downloading malicious content or pornographic and violent content. Thus, we cannot publish an app using this plug-in technique on Google Play.

This book will teach you the underlying knowledge of the Android system, which help you master Android technique at a high level.

After reading this book, you will be familiar with several aspects of the Android system, including the following content:

- *Binder* and *AIDL* mechanisms.

- The working mechanisms of *Activity, Service, ContentProvider*, and *BroadcastReceiver*.

- Communication between *ActivityManagerService* and four components.

- How to launch an app.

- *LaunchMode*.

- The families of *Context* and *ClassLoader*.

- *MultiDex.*

- How to load SO.

- *PackageManagerService* and how to install an app in the Android system.

- Reflection.

- The *Proxy.newProxyInstance()* method for adding an external function to the original API.

WHO IS THIS BOOK FOR?

Don't use the plug-in techniques introduced in this book on Google Play; it's forbidden.

This book introduces a lot of knowledge about the Android system which is useful to app developers.

THE SAMPLE CODE

The sample code in this book is on my Github: https://github.com/BaoBaoJianqiang/.

There are 74 demos in this book, and I list the address of each demo in the corresponding chapter and section.

In Appendix A, I list all the demos with their corresponding chapter and section.

THE BOOK'S STRUCTURE

This book consists of 11 chapters: chapters 1 to 5 introduces the basic knowledge of plug-in techniques; chapters 6 to 10 introduces several solutions for plug-in programming issues; Chapter 11 is an overall summary.

Below is a brief overview of the chapters.

Chapter 1 introduces the history of Android plug-in techniques.

Chapter 2 introduces the underlying Android system, including *Binder* and *AIDL*, *ActivityManagerService*, *PackageManagerService*, *ActivityThread*, *LoadedApk*, and so on. As this book is written for app developers, I illustrate these concepts with a series of pictures rather than code.

Chapter 3 introduces the syntax of reflection, and the encapsulation of the reflection, including jOOR, a famous Java reflection framework. Reflection is the basis of plug-in techniques.

Chapter 4 introduces proxy patterns, including *Static-Proxy* and *Dynamic-Proxy*, these two proxies generate two important plug-in frameworks, *DroidPlugin* and *DL*. *Proxy.newProxyInstance()* is a widely used plug-in, and we use this method to hook *IActivityManager* and *IPackageManager* in this chapter.

Chapter 5 introduces how to start an *Activity* not declared in the *AndroidManifest.xml*, based on the *Proxy.newProxyInstance()* introduced in Chapter 4.

Chapter 6 introduces the basic knowledge of plug-ins, including how to debug from the Hostapp to the plug-in, and interface-oriented programming.

Chapter 7 introduces how to load *Resources* into plug-ins. *AssetManager* and *Resources* are key points, especially the method *addAssetPath()* of *AssetManager*. Based on this technique, we can dynamically change a skin.

Chapter 8 introduces plug-in solutions for *Activity*, *Service*, *BroadcastReceiver*, and *ContentProvider*. A different mechanism of these four components results in different plug-in solutions.

Chapter 9 introduces a plug-in framework based on *Static-Proxy*. The creator of this framework invented a new keyword "that," so this framework is also called "That." "That" is a very smart framework; it's also well known as Puppet.

Chapter 10 considers other related plug-in techniques. Including how to resolve conflicts between the resource ID in plug-ins, how to use fragments in plug-ins, how to replace HTML5 with *Activity*, how to use *ProGuard* in plug-ins, how to reduce the size of plug-ins, how to download a SO file dynamically, and how to support the Android O and P systems with plug-ins.

Chapter 11 summarizes all the plug-in techniques mentioned in this book.

CONTACTING THE AUTHORS

If you have suggestions, remarks, or questions on plug-in techniques and sample code, please contact the author on: 16230091@qq.com.

CHAPTER 1

Plug-Ins from the Past to the Future

G OOGLE PLAY HAS A strict app auditing strategy. For example, it forbids any app to download another app to prevent it from downloading bad content, pornographic and violent content, for example.

In addition, Google Play forbids app developers from modifying the original behavior of the API of the Android system, which is not open to the app developers. For example, the method *addAssetPath()* of the *AssetManager*, and the method *currentActivityThread()* of the *ActivityThread*. Also, Android P launched a new mechanism named the grey-list and black-list. If the developer modifies the APIs through the two lists above, these APIs will print a warning or return *null* directly.

The auditing strategy in the Chinese app market is less strict. Downloading and launching are allowed and there are two main techniques widely used in China; one is plug-in, the other one is hot-fix.

1.1 ANDROID PLUG-INS IN CHINA

The plug-in technique separates one app into a lot of small apps for different business purposes; the OTA* app, for example, consists of hotels, flights, cars, and other domains. We can separate these domains into several small apps, such as a hotel app, flight app, and car app, and all these small apps are called plug-in apps. As all the businesses are separated into

---

* OTA: Online Travel Agent

different plug-in apps, only the home page is left in the main app (and is called the Hostapp). When users click the button in the Hostapp, it will navigate to the small apps.

In traditional app coding strategy, all the code and logic should be in one app. When we find bugs in the app, there is only one way to solve this online bug; it is to package this app again and submit it to the Android app market. However, the users must download the latest version of this app to remedy the bugs. This is not a good solution; it's not user-friendly. Most users don't want to waste time updating an app.

Android plug-ins are a very good solution to the problem above. If there is a bug in one plug-in, we just need to package this plug-in app again, and then put this new plug-in on the remote server. When the app user opens the Hostapp, it will download this new plug-in in the background thread automatically. When downloaded successfully, the user needs to restart the app and the bugs will have been eliminated from the app.

The plug-in technique is not only used to fix bugs but is also suitable for rapid software development. In traditional app development, you launch a new version of the app every month. It is very common for a very big company to have 100 product requirements needing to be coded within one month. Any delay in development causes some products to launch later than planned. By using the plug-in technique, the different apps can be launched individually; there will be no time limit.

In China, the hot-fix technique was developed using similar ideas to plug-ins. The hot-fix technique is useful for fixing online bugs. When app developers find online bugs, they can fix the codes and then package the code modification into a zip file; then upload this zip file to the remote server, so that users can download this zip file dynamically. After the users have downloaded the zip successfully, the app will decompose this zip file and substitute the old code with the new code in the zip file.

The plug-in technique and hot-fix technique were developed using very similar ideas. The plug-in technique loads outside the apps by hooking the Android internal system API. The hook occurs in the Java code, where the hot-fix is happening in NDK, which means C++. A hot-fix replaces the pointer of the old method with a pointer of the new method.

This book focuses on the plug-in technique.

1.2 HISTORY OF ANDROID PLUG-IN TECHNIQUES

On July 27, 2012, the first milestone in Android plug-in technology was reached. Yimin Tu, who worked for Dianping.com, released the first

Android plug-in open source project, *AndroidDynamicLoader**, and the Dianping.com app was based on this framework. This plug-in framework is based on Fragment. The app has only one activity; all the pages are implemented by fragments and loaded by this activity. Some fragment pages are plug-ins, which can be downloaded dynamically. This plug-in framework was the first time anyone used the method *addAssetPath()* of the *AssetManager* to handle the *Resources* in the plug-in app.

In 2013, 23Code appeared. 23Code provides a container where plug-ins can be dynamically downloaded and run. We can write a variety of UserControls and run them in 23Code. It is an Android plug-in framework, but without source code and not widely known.

On March 27, 2013, Bokui, the developer of the Taobao app, shared technical information on Taobao's plug-in framework. The name of this plug-in framework is Atlas†. In this topic, he introduced a way to modify the internal API of Android, incremental update, downgrade, compatibility, and so on. It's a pity that this plug-in framework is not an open source project. We can't learn more from this topic.

At 8:20 on March 30, 2014, the second milestone of Android plug-in technology was reached. Yugang Ren published an Android plug-in project named *dynamic-load-apk‡*, which was not the same as the other plug-in projects. It did not modify the internal methods of the Android system, but solved problems from the application layer of the app by creating a class named *ProxyActivity* to distribute and start the activity of the plug-in. Yugang Ren invented a keyword called "that" in this framework, it's also called the "That" framework in this book. In fact, the creator does not like this nickname and named it *DL* for short. When he developed this framework, there were so many difficulties, because there was not enough information on Android plug-in technology that could be referred to, especially before 2014.

The "That" framework only has the implementation of *Activity* at the beginning. With the contribution of Xiao Tian and Siyu Song, the implementation of *Service* was available later. In April 2015, the "That" framework was stable.

At the same time, Tao Zhang was also contributing to the implementation of plug-in technology. In May 2014, after reading all the source codes of *DL*,

---

* https://github.com/mmin18/AndroidDynamicLoader
† http://v.youku.com/v_show/id_XNTMzMjYzMzM2.html
‡ https://github.com/singwhatiwanna/dynamic-load-apk

he released his first plug-in framework, *CJFrameForAndroid**. This design was similar to the "That" framework. In addition, the *CJFrameForAndroid* framework provided a plug-in solution called *LaunchMode*, which was a very important contribution to plug-in techniques.

In November 2014, Houkx released a plug-in project named *android-pluginmgr* on GitHub†. This framework first proposed registering a *StubActivity* in the *AndroidManifest.xml* to cheat the *AMS* but opened an *Activity* in a plug-in. At the same time, Houkx also found that all the permissions should be declared in the *AndroidManifest.xml* of the plug-in in advance.

On December 8, 2014, there was good news, Android Studio V1.0 was available. Android developers began to gradually abandon Eclipse to use Android Studio. Android Studio is compiled and packaged with Gradle, which makes the design of plug-in frameworks much simpler, eliminating the inconvenience of using Eclipse to run the Android SDK.

Then, though, 2015 was coming. Lody, an 18-year-old boy, began using Android in 2015 when he was a senior high school student. He had studied the source code of the Android system for 3 years. His first well-known open source project was *TurboDex‡*, which could quickly load *dex* with high speed. This is a very useful plug-in framework because it usually takes a long time to load all the plug-ins for initialization.

At the end of March 2015, Lody released the plug-in project: *Direct-Load-Apk§*. This framework combined two thoughts mentioned earlier; one was *Static-Proxy*, from Yugang Ren's "That" framework, the other one was to cheat the *AMS*, from Houkx's *pluginmgr* framework. *Direct-Load-Apk* is not widely known, because Lody had too much school homework.

The legend of Lody wasn't finished yet; he spent a lot of time on *VirtualApp*. *VirtualApp* is like a virtual machine on the Android system. It can install and run other apps. We'll discuss *VirtualApp* in Section 1.6.

In May 2015, Limpoxe released the plug-in framework: *Android-Plug-in-Framework¶*.

---

* https://github.com/kymjs/CJFrameForAndroid
† https://github.com/houkx/android-pluginmgr
‡ https://github.com/asLody/TurboDex
§ http://git.oschina.net/oycocean/Direct-Load-apk
¶ https://github.com/limpoxe/Android-Plugin-Framework

In July 2015, Kaedea released the plug-in framework: *Android-dynamical-loading**.

On August 27, 2015, the third milestone of Android plug-in technology was reached, Yong Zhang's plug-in framework *DroidPlugin*[†] came out. Yong Zhang was a developer at Qihoo360, and *DroidPlugin* was used in his team. The magic of this framework is that any app can be loaded into the HostApp. You can write a HostApp based on this framework, and then load apps written by others as plug-ins.

DroidPlugin is a powerful plug-in framework, but its disadvantage is obvious. It modifies too many internal APIs of the Android system. Due to the lack of literature on the *DroidPlugin* framework, it is difficult to understand. There are many articles about *DroidPlugin* on blogs and forums, but the best one is written by WeiShu Tian[‡]. He also worked at Qihoo360 and had the opportunity to talk about *DroidPlugin* with its creator. He then wrote a series of articles about the *DroidPlugin*, including the principles of *Binder*, *AIDL* and the plug-in mechanism for the *Activity*, *Service*, *BroadcastReceiver*, and *ContentProvider*.

The year 2015 was the first year of Android plug-in development. Not only the "*that*" framework and *DroidPlugin* but many other plug-in frameworks were also born at that time.

The project *OpenAtlas* was released on GitHub in May and was later renamed *ACDD*. It proposes modifying the command aapt so that the resource ID of the plug-in is no longer a fixed value of 0x7f, but can be modified to a value such as 0x71. aapt is a command line tool supplied by Android. It's used to generate resource IDs during the packaging process of an Android app. This technique solves the problem of resource ID conflict after merging the resources of the plug-in and the Hostapp together.

OpenAtlas hooks the method *execStartActivity()* of *Instrumentation* to load the *Activity* of the plug-in dynamically.

In addition, *OpenAtlas* also modifies *ContextWrapper*, and rewrites the method *getResources()*. Because the *Activity* is the subclass of the *ContextWrapper*, the *Activity* of the plug-in inherits the method *getResources()* to get the resources of the plug-in. We can't use this method anymore; we need to create a parent class named *BasePluginActivity*, all the

---

* https://github.com/kaedea/android-dynamical-loading
† https://github.com/Qihoo360/DroidPlugin
‡ Blog address: http://weishu.me

Activities of the plug-in should inherit *BasePluginActivity* and override the method *getResources()* to fetch the resources of the plug-in.

Ctrip.com released their plug-in framework *DynamicAPK** in October 2015, which was based on the *OpenAtlas* framework.

At the end of December 2015, Guangliang Lin published his plug-in framework, *Small†*. At that time, he worked for a car trading platform and his framework was used for this platform.

Small has a lot of interesting features:

- Merges all *dex* files of the plug-in into a *dex* array of the Hostapp. This means the HostApp can load any class of plug-in.

- Declares *StubActivity* in the *AndroidManifest.xml* to cheat the *AMS*. This solution is the same as the *DroidPlugin*.

- Pre-declares *Service, Receiver,* and *ContentProvider* of the plug-in in the *AndroidManifest.xml* of the Hostapp to support the other three components.

- Invokes the method *addAssetPath()* of the *AssetManager,* and all the resources of the plug-in are merged into the resources of the Hostapp. If a resource ID conflict occurs, *Small* modifies *R.java* and *resource. arsc* in the packaging process. After these two files are generated, *Small* will change the resource ID in these two files; for example, from 0x7f to 0x71.

At the end of 2015, all the technical issues of plug-ins were resolved. That year, the plug-in technology was very varied. A lot of open source plug-in frameworks were born in 2015. These frameworks were almost all invented by individuals. It was basically divided into two categories. Dynamic-Hook was invented by Yong Zhang in DroidPlugin, and Static-Proxy was invented by Yugang Ren in the "That" framework.

In 2015, Android hot-fix technology and *React Native* appeared in the app developer's world, which provided the same advantages as Android plug-in technology. Android plug-in technology was no longer the only choice for app developers.

---

* https://github.com/CtripMobile/DynamicAPK
† https://github.com/wequick/Small

Since 2016, many internet companies have released their plug-in frameworks. These frameworks focus on stability and compatibility to benefit their millions of daily users.

Let us enumerate these frameworks in order:

- In August 2016, iReader published its plug-in framework named *ZeusPlugin**.

- In March 2017, Alibaba published its plug-in framework named *Atlas*†.

In June 26, 2017, Qihoo360 published its second plug-in framework named *RePlugin*‡, which is different from *DroidPlugin*.

On June 29, 2017, Didi published its plug-in framework named *VirtualApk*§.

All the plug-in frameworks published on github by internet companies focus on:

- Plug-in compatibility, including the impact of the upgrade of the Android system on the plug-in framework, and the impact of different mobile phone ROMs.

- Plug-in stability, for example, crashes.

- Management of plug-ins, including installation and uninstallation.

In spite of the fact that a few years have passed, various plug-in frameworks have gradually become stable. Developers who are now engaged in plug-in technology only need to pay attention to the annual upgrade of the Android system and add code to be compatible with the latest updates.

With the maturity of the plug-in framework, many authors of plug-in technology have begun to change their focus. Some of them are still sticking with Android; for example, Yugang Ren is still working in the Android domain; and some of them have moved to Blockchain where they focus on writing smart contracts with the Go language every day.

---

* https://github.com/iReaderAndroid/ZeusPlugin
† https://github.com/alibaba/atlas
‡ https://github.com/Qihoo360/RePlugin
§ https://github.com/didi/VirtualAPK

Thanks to those who have contributed to plug-in technology, including the authors of plug-in frameworks, as well as the authors who have written a series of articles to introduce this technique.

1.3 USAGE OF PLUG-INS

Once upon a time, we naively thought that Android plug-ins were intended to add new features.

It took a lot of time and effort, but after the project was implemented with plug-in frameworks, we discovered that 80% of the usage of plug-ins was to fix online bugs. At this point, it has the same capabilities as hot-fix techniques such as *AndFix** and *Tinker†*, and even better than these hot-fix frameworks.

The app always releases a new version every two weeks. Generally, the new feature goes online at this point. On the other hand, in some companies, the release strategy of Android app is affected by iOS app releases in the App Store.

In the time before plug-in frameworks, we were scared to write bugs; if a serious bug appeared, we had to release a new version to fix this bug. The app users would have to update the app to the newest version, and it's not user-friendly to have to make frequent downloads to keep the app running smoothly.

With the plug-in framework, developers don't have to worry about the quality of the code—if something is wrong, you can release a new version to fix it quickly. After the app is released, each plug-in will have one or two new releases every day.

The Android plug-in framework has become a bug-fixing tool and this is something that we don't want to see.

In fact, the plug-in framework is more suitable for MMORPG games. There will always be new skins, or a new hero role available every few days, even for adjusting the attributes. All of these do not need to be released as a new version.

There is another use for plug-in technology, which is the ABTest, but it is not used widely. When the product manager wants to determine which will be selected from two styles of design, there will be two strategies which are made into the two plug-in packages; 50% of the users download strategy A, the other 50% download strategy B. Checking the results after

---

* https://github.com/alibaba/AndFix
† https://github.com/Tencent/tinker

a week, such as the page conversion rate, will tell you which strategy is better. This is called a data-driven product.

In the previous section, the componentization of Android was mentioned. That is, with the independence of the business unit, the Android and iOS teams are split into their own business and have their own organization relationships. Therefore, it is necessary to split the different services of hotels and flights into different modules. In the componentization of Android, modules are still dependent on the *aar* file; we can use Maven to manage these *aar* files.

The componentized model of Android is only applicable to the development stage. Once there is a bug in the online release, or new features are to be released, all modules must be packaged together again to deploy the new release.

Plug-in technology is the final solution based on Android componentization. At this point, each service module is no longer an *aar* file, but an *apk* file, placed in the folder *assets* of the Hostapp. In this way, after a release, some modules are updated, only the code of this module is packaged again, an incremental package is generated, and it is put on the remote server for the app users to download.

1.4 ANOTHER CHOICE: REACT NATIVE

In 2015, *React Native* was born. At that time, few people paid attention to it because it was still immature with only a few basic functions. Then, with the next iteration of the *React Native* project, the functions were much improved. Although there has not been a release version 1.0 so far, we find that it is a perfect "plug-in" framework to support both Android and iOS systems.

React Native is written based on JavaScript, packaged, and put on the remote server for Android and iOS apps to download and use.

For small-sized or medium-sized companies and startups, who don't have the human and financial resources to develop a plug-in framework, generally adopt a relatively stable, open source, and continuous updated plug-in framework. However, it seems that iOS has no technical framework, especially after the jsPatch (a hot-fix solution) was forbidden by the App Store. Their best choice is *React Native*. Once the JavaScript is recruited, it will be able to quickly iterate and release.

On *React Native*, there are already some books available. This book mainly introduces Android plug-in technology. This section shows some points that Android plug-in is not as good a technique as *React Native*.

1.5 DO ALL COMPONENTS REQUIRE PLUG-INS?

In Android, *Activity, Service, ContentProvider*, and *BroadcastReceiver* are the four major components.

Do all four components need to convert to plug-ins? Over the years, I have been working on plug-in technology with this question.

I have worked in OTA companies for several years. This kind of app is similar to e-commerce ones, including a complete set of payment processes, and *Activity* is the most used; 200 or 300 *Activities* are not surprising. The other three components are rarely used.

Most apps in China have the same situation.

According to the technology stack, the app is divided into four domains:

- Game App. People have their own online update process. Many of them use scripts like Lua.

- Mobile assistants, mobile phone guards, and the use of such applications for *Service, Receiver*, and *ContentProvider*.

- Music, video, and live video applications are very dependent on *Service* and *Receiver*, in addition to more *Activities*.

- E-Commerce, social, news, and reading apps use a lot of *Activities*. The use of the other three major components is not enough.

Different plug-in frameworks are suitable for different requirements. If an app uses a lot of the *Service, BroadcastReciever*, and *ContentProvider* components, the plug-in framework for this app must support all four components in Android, but this plug-in framework is hard to maintain. Otherwise, a plug-in framework which only supports *Activity* is enough for a simple app.

1.6 DOUBLE-OPENING AND VIRTUAL MACHINE

Since plug-ins will be replaced by *React Native* in the following years, what is the future of plug-ins? The answer is virtual machine technology.

Some engineers already have experience of installing a virtual machine on a computer. If the computer's memory is large enough, you can open multiple virtual machines at the same time. On each virtual machine, you can log into Twitter with a different account, and can then chat with yourself. Of course, chatting with yourself doesn't make any sense.

Can we support installing one or more virtual machines on an Android system? Lody, a college student, is doing such work. He has a very famous open source project, VirtualApp, which is now in commercial operation*.

With such a virtual machine system, we can open two Twitter apps with a different account in only one mobile phone and chat with ourselves.

The multiple instances of technology that opens an app at the same time is called double-opening. Some mobile phone systems in China now support double-opening technology. You can see this option in the settings of Android phones.

The technology of double-opening and virtual machines is outside of this book's scope.

1.7 FROM NATIVE TO HTML5

Through the immense prosperity of app technology, I got my first job in 2004; when the computer assembly industry was transitioning from *CS (Client/Server)* to *BS (Browser/Server)* architecture. For example, if you installed MSN on your computer, you can chat with your friends through this software. After the technology of the internet grew up, you started to move the original system into the website. This is *BS* architecture.

Compared to *CS*, *BS* was a thin client; many features were not supported by *BS*. Designers came up with the concept of a SmartClient, which is a CS pattern. Outlook is a good sample. You can read and write email offline without a network; it sends a written email automatically.

Then, Flash became hot. Flash was the originator of the web-rich client. Based on Flash, there is Flex, which is now being adopted by more and more companies. At this time, Microsoft also created Silverlight, which is like Flash but on the web. At the same time, JavaScript was also working hard and became the final winner in the rich client space. A book on JavaScript was very popular at that time; it is called "JavaScript Design Patterns."

JavaScript was only used for web visual effects in 2004. From 2005 onward, JavaScript has experienced additions with *Ajax*, *jQuery*, *ECMAScript* 1 to 6, and so on. The frameworks of *Angular*, *React*, and *Vue* have become extremely powerful and have been packaged due to JavaScript as an "object-oriented" language.

---

* https://github.com/asLody/VirtualApp

Compared to web technology, the app is taking the same development path. In the first five years of mobile development, app developers used basic syntax to write each app. The experience was not user-friendly. The networking speed was not fast and many white screens were found in the apps, and there were a lot of bugs and crashes. Over the following five year of mobile development, however, more and more mobile techniques were invented by app developers, such as *RxJava*, *ButterKnife*, *jsPatch*, *Tinker*, and a lot of plug-in frameworks. The next stage is the transition from *CS* to *BS*. The hybrid technology is the above-mentioned *BS*, but there are many defects, especially the bad performance of web browsers. Then there is *React Native*. HTML5 is also slow, but you can translate HTML5 into *Native* code. I don't know which techniques will appear in the next five years, but the trend is obvious. Android and iOS technology will not die; on the other hand, HTML5 will become the main method for app development in the coming years.

1.8 SUMMARY

In this chapter, we reviewed the history of Android plug-in technology, which is basically divided into two parts, *Dynamic-Proxy* and *Static-Proxy*. All plug-in frameworks are based on these two parts. After the history review, we found that plug-in technology was not accomplished at one stroke, but it has experienced a process of gradual improvement.

Plug-in technology is not only used to fix bugs and dynamically release new features. In the process of researching plug-in technology, we have developed the Android virtual machine and double-opening technology. This is a new technology area that can bypass Android system limitations and run the app faster.

React Native was also mentioned, which also fixes bugs and dynamically releases new features, just like Android plug-in technology. Which technology is selected depends on whether the R&D team is based on HTML5 or Android and depends on whether it will be released on Google Play.

CHAPTER 2

The Underlying Knowledge of Android

THIS CHAPTER IS BASED on a series of articles I wrote in 2017: "Underlying Android knowledge for app developers." It's necessary to master this knowledge before we study plug-in technology.

2.1 OVERVIEW OF UNDERLYING ANDROID KNOWLEDGE

As early as when I was a junior Android developer, I was puzzled about a lot of concepts, and I could not find answers to questions such as "how to install apks in Android" or "how to start an *Activity* in Android." On the other hand, AIDL is mentioned in many books, but I haven't used this technique in my apps. The Android system has four important components declared in the *AndroidManifest.xml*, i.e. *Activity, ContentProvider, BroadcastReceiver,* and *Service*, but I mostly use Activity. I haven't used any of the other three components, like *ContentProvider*. I can currently only study these three components in Android development books.

Because I am working in an OTA company, I am familiar with app development. For this kind of app, it basically consists of a list page and a detail page. *Activity* is widely used in this app; what we need to do is to capsulate the complex network communications.

Almost all app developers have the same experiences.

There are many books that introduce the underlying knowledge of Android, but they are written for the Android ROM developers. They are not suitable for app developers to read and study.

So, for the past few years, I have been looking for such a kind of knowledge, which helps app developers write code more efficiently but spend less time studying the underlying code in the Android system.

There are two types of knowledge:

- Type 1: knowing the concept but not needing to master it in detail, such as *Zygote*, *SurfaceFlinger*, and *WMS*. In fact, app developers needn't study *Zygote*. It is enough to know that *Zygote* is the entry point of the Android system.

- Type 2: knowing the principles but not needing to read its implementation, such as with *Binder*. There are thousands of hundreds of articles that introduce *Binder*, but these articles are not suitable for app developers. App developers should master the architecture of *Binder*, such as *Binder Client*, *Binder Server*, and *ServiceManager*, but don't spend time studying code implementation.

Communication between the *AMS* and the four components, such as *Activity*, is based on a *Binder*. App developers should master which classes act as a *Binder Server* and which classes act as a *Binder Client*. Sometimes *AMS* acts as a *Binder Server* and *Activity* acts as a *Binder Client*; sometimes *Activity* acts as a *Binder Server* and *AMS* acts as a *Binder Client*. It's useful for app developers to learn *AIDL*.

I will introduce the following concepts in this chapter.

- *Binder*

- *AIDL*

- *ActivityManagerService*

- The working Principle of *Activity*, *Service*, *ContentProvider*, and *BroadcastReceiver*

- *PackageManagerService*

- The installation process of an app

- *ClassLoader* and *Parent-Delegation*

- The family of *Context*

2.2 BINDER

Binder is used for cross-process communication.

There are too many articles about *Binder,* and each article introduces knowledge from the Java layer to the C++ layer. But this is not helpful for app developers. I want to introduce *Binder* without code implementation in this section.

1) *Binder* is divided into two processes, *Client* and *Server.*

Client and *Server* are relative. Whoever sends the message will be the *Client.* Whoever receives the message will be the *Server.*

For example, two processes A and B; they use *Binder* to communicate with each other, A sends a message to B, at this time A is the *Binder Client,* B is the *Binder Server;* and then B sends a message to A, at this time B is the *Binder Client,* A is the *Binder Server.*

2) Figure 2.1 shows the basic architecture of *Binder.*

In Figure 2.1, *ServiceManager* is a container, and *Binder Server* and *Binder Client* are both registered in this container. *Binder Driver* is responsible for the communication between *Binder Server* and *Binder Client.*

ServiceManager is like a telephone exchange. A telephone exchange has an address book to store the telephone numbers for each person. When

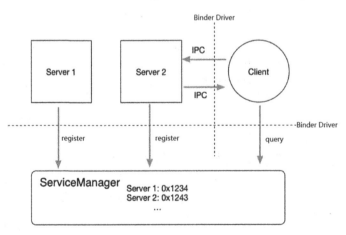

FIGURE 2.1   Structure of *Binder.*

Tom wants to call Jerry, Tom will dial up Jerry's phone number. The signal will be transferred to the telephone exchange. The operator working in the telephone exchange takes the role of a *Binder Driver*; he will search this telephone number in the address book. Because Jerry's telephone number is registered in the telephone exchange, Jerry's phone is called; without registration, the telephone exchange would tell Tom that the phone number does not exist.

In contrast to the Android *Binder* mechanism, Tom is the *Binder Client*, Jerry is the *Binder Server*, and the telephone exchange is the *ServiceManager*. The operator working in the telephone exchange is busy working on this process; it's the *Binder Driver* in Figure 2.1.

3) Let's have a look at the process of *Binder* communication in Figure 2.2.

The *Binder Client* can't invoke the method *add()* of the *Binder Server* directly. Because they are in different processes, we can use *Binder* to finish this work. *Binder Driver* will create a new class *Proxy* to as the wrapper of *Binder Server*. *Binder Client* can communicate with *Proxy*.

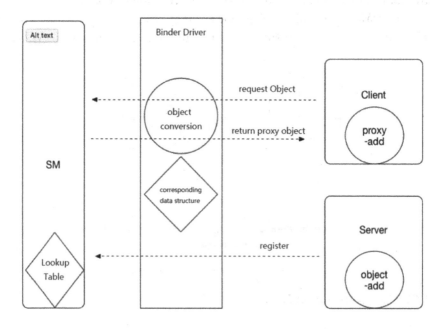

FIGURE 2.2 Communication in *Binder*.

The communication has the following three steps; we use *Server* and *Client* instead of *Binder Server* and *Binder Client* for short.

1) The server is registered in *ServiceManager*.

2) *Binder Driver* generates a class *Proxy* for the *Server*. *Proxy* is the wrapper of the *Server*.

3) The client invokes the method *add()* of *Proxy*; it will invoke the method *add()* of the *Server* indirectly.

Binder Driver is working hard. But the app developers needn't learn the underlying implementation of the *Binder Driver*.

Up until now, I have introduced the knowledge of *Binder*. Based on this knowledge, we go on to study *AIDL* and the *AMS*.

2.3 AIDL

AIDL is an extension of *Binder*. *AIDL* is widely used in the Android system, such as in the *Clipboard*.

We need to know the following classes in *AIDL*:

- *IBinder*

- *IInterface*

- *Binder*

- *Proxy*

- *Stub*

When we write an *AIDL* (such as *MyAidl.aidl*, which has a method *sum()*, Android Studio will help us generate a class named *MyAidl.java*, shown in Figure 2.3.

We split the class *MyAidl* into three files. It's easy to read, and shown as follows:

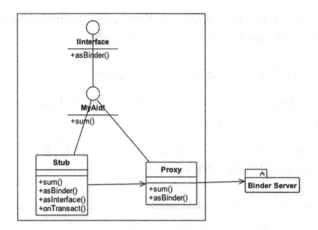

FIGURE 2.3   Class diagram of *AIDL*.

```java
public interface MyAidl extends android.os.IInterface {
  public int sum(int a, int b) throws android.
  os.RemoteException;
}

public abstract class Stub extends android.os.Binder
implements jianqiang.com.hostapp.MyAidl {
  private static final java.lang.String DESCRIPTOR =
  "jianqiang.com.hostapp.MyAidl";

  static final int TRANSACTION_sum = (android.
  os.IBinder.FIRST_CALL_TRANSACTION + 0);

  /**
  * Construct the stub at attach it to the interface.
  */
  public Stub() {
    this.attachInterface(this, DESCRIPTOR);
  }

  /**
  * Cast an IBinder object into an jianqiang.com.
  hostapp.MyAidl interface,
  * generating a proxy if needed.
  */
  public static jianqiang.com.hostapp.MyAidl
  asInterface(android.os.IBinder obj) {
```

```
if ((obj == null)) {
  return null;
}
android.os.IInterface iin = obj.queryLocalInterface
(DESCRIPTOR);
if (((iin != null) && (iin instanceof jianqiang.
com.hostapp.MyAidl))) {
  return ((jianqiang.com.hostapp.MyAidl) iin);
}
return new jianqiang.com.hostapp.MyAidl.Stub.
Proxy(obj);
}

@Override
public android.os.IBinder asBinder() {
  return this;
}

@Override
public boolean onTransact(int code, android.
os.Parcel data, android.os.Parcel reply, int flags)
throws android.os.RemoteException {
  switch (code) {
    case INTERFACE_TRANSACTION: {
      reply.writeString(DESCRIPTOR);
      return true;
    }
    case TRANSACTION_sum: {
      data.enforceInterface(DESCRIPTOR);
      int _arg0;
      _arg0 = data.readInt();
      int _arg1;
      _arg1 = data.readInt();
      int _result = this.sum(_arg0, _arg1);
      reply.writeNoException();
      reply.writeInt(_result);
      return true;
    }
  }

  return super.onTransact(code, data, reply, flags);
}
}
```

```
class Proxy implements jianqiang.com.hostapp.MyAidl {
  private android.os.IBinder mRemote;

  Proxy(android.os.IBinder remote) {
    mRemote = remote;
  }

  @Override
  public android.os.IBinder asBinder() {
    return mRemote;
  }

  public java.lang.String getInterfaceDescriptor() {
    return DESCRIPTOR;
  }

  @Override
  public int sum(int a, int b) throws android.
  os.RemoteException {
    android.os.Parcel _data = android.os.Parcel.
    obtain();
    android.os.Parcel _reply = android.os.Parcel.
    obtain();
    int _result;
    try {
      _data.writeInterfaceToken(DESCRIPTOR);
      _data.writeInt(a);
      _data.writeInt(b);
      mRemote.transact(Stub.TRANSACTION_sum, _data,
      _reply, 0);
      _reply.readException();
      _result = _reply.readInt();
    } finally {
      _reply.recycle();
      _data.recycle();
    }
    return _result;
  }
}
```

There are one interface and two classes in *MyAidl.java*, *Stub* and *Proxy* which implement the interface in *MyAidl*. *Proxy* is the inner class of *Stub*.

Stub has two important methods, *asInterface()* and *onTransact()*. In fact, Figure 2.3 is not the full picture of *AIDL*; it's only a *Binder Client*. In Figure 2.4, we can find both a *Binder Server* and *Binder Client*.

In Figure 2.4, *Stub* in the left rectangle is the *Binder Client*; *Stub* in the right rectangle is the *Binder Server*. These two *Stubs* can't communicate with each other. They use *Proxy* to communicate with each other.

Let's analyze this process step by step.

1) Look at the *Binder Client* in Figure 2.4. We always write code as follows:

```
MyAidl.Stub.asInterface(xxx).sum(1, 2);    ///xxx is an IBinder
                                              object
```

The method *asInterface()* of *Stub* is used to judge if the *IBinder* object is in the current process.

- If yes, convert this *IBinder* object and return directly.

- If no, return a *Proxy* object. *Proxy* is the wrapper of *Stub*. To invoke the method *sum()* of *Stub*, will invoke the method *sum()* of *Proxy* indirectly. The code is as follows:

```
return new MyAidl.Stub.Proxy(obj);
```

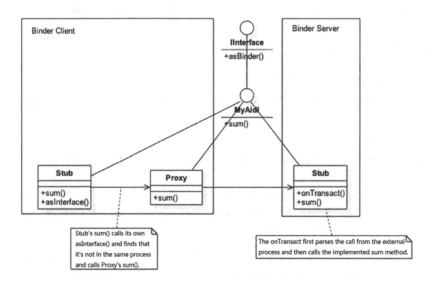

FIGURE 2.4   Full picture of *AIDL*.

2) The method *sum()* of *Proxy* prepares data, writes the name and parameters of the method *sum()* to the variable *_data*, and sends *_data* to the *Binder Server*. It also declares a variable *_reply* to receive the return value from the *Binder Server*, as shown in the code above. *mRemote* is the parameter of the method *asInterface()*; it's an *IBinder* object. The method *transact()* of *IBinder* is used to send data to the *Binder Server*:

```
mRemote.transact(Stub.TRANSACTION_addBook, _data,
_reply, 0);
```

3) Look at the *Binder Server* in Figure 2.4. *Stub* uses the method *onTransact()* to receive data from the *Binder Client*, including the name and parameters of the method. *Stub* will find the corresponding method in the current process, execute this method, and return the value to the *Binder Client*. All this logic is written in the method *onTransact()* of *Stub*.

In Section 2.5, 2.8, 2.9, and 2.10, I will introduce the four components, *Activity*, *Service*, *BroadcastReceiver*, and *ContentProvider*. The communication between these components and the *ActivityManagerService* (*AMS* for short) is based on *Binder*. When *Activity* sends a message to the *AMS*, *AMS* is the *Binder Server* and *Activity* is the *Binder Client*. When the *AMS* notifies an *Activity* to do something; the roles are reversed.

For the task *Activity*, for example, we may have a question about which class corresponds to *Stub* and which class corresponds to *Proxy*. I will introduce this topic in Section 2.5.

2.4 ACTIVITYMANAGERSERVICE

It so far seems that the *AMS* only communicates with *Activity*, but *AMS* also communicates with the other three components. Don't be confused by its name.

Before writing this book, I had two questions for a long time:

1) During the app installation process, why not extract the *apk* file into the phone, As it's easier to read the *apk* files directly? I will answer this question in Section 2.11.

2) There are two processes when *AMS* communicates with the four components. One is the system process which hosts *AMS* logic, the

other one is the app process which hosts *Activity*, written by the app developer. We could hook logic in the app process, but we can't hook the logic in the system process. Why?

The *AMS* is one of the services that runs during Android system processes. After we learn a bit about the *AMS*, we find it's easy to understand. Take the *Clipboard* in Android as an example. If we write some code in our own app to modify the original function of Clipboard in the Android system process, it will have effects on the other apps installed on the system. It's a virus! So, Android forbids the app from modifying system services such as the *AMS*. We can only modify the behavior of *Clipboard* in the current app.

2.5 *ACTIVITY* WORKING PRINCIPLES

Activity is widely used in apps. In this section, I will talk about how to launch an app.

2.5.1 How to Launch an App

Click an app icon in the *Launcher* of an Android system, for example, the Amazon App, and we can see the home page (or guide page) of this app. The action is simple, but the communication between *Activity* and the *AMS* is very complex.

The *Launcher* is an app supplied by the Android system.

When we write an app, we must set the default *Activity* which is first launched in the *AndroidManifest.xml*, shown as follows:

```
<activity android:name=".MainActivity">
   <intent-filter>
      <action android:name="android.intent.action.
      MAIN" />
      <category android:name="android.intent.
      category.LAUNCHER" />
   </intent-filter>
</activity>
```

When we install this app on a mobile phone, the *PackageManagerService* will parse the *AndroidManifest.xml* and store three pieces of information in the *Launcher*. Take Amazon as an example:

```
action:android.intent.action.MAIN
category: android.intent.category.LAUNCHER
cmp: package name of Amazon + first start Activity
```

When we click the app icon in the *Launcher*, it will find these three pieces of information and start the first *Activity* of the Amazon app.

2.5.2 Starting the App Is Not So Simple

In Section 2.5.1, we gave a quick overview of the process of starting an app. But we find *Launcher* and Amazon are two different apps; they use different processes, and they can't communicate with each other directly. With the help of *Binder*, one app sends a message to the *AMS*, and the *AMS* notifies the other app.

Take the Amazon app as an example; the whole process is as follows*:

1) When the user clicks on the app icon in *Launcher*, *Launcher* informs the *AMS*, "I want to open Amazon, I will send an action and a category, please help me open the *Activity* which is suitable for this information."

2) The *AMS* informs the *Launcher*, "Your work is done." At the same time, the *AMS* records which activity to start.

3) The current *Activity* of the *Launcher* invokes its method *onPause()*, and informs the *AMS*, "I'm going to sleep, you can launch the Amazon app."

4) The *AMS* checks whether the Amazon app has already been launched. If yes, it wakes up the Amazon app; otherwise, the *AMS* must start a new process for the Amazon app. The *AMS* creates an *ActivityThread* object in the new process and executes the function *main()* to launch the Amazon app.

5) After the Amazon app starts, it will inform the *AMS*, "Hey, I am the Amazon app, I'm ready to start."

6) The *AMS* fetches the action and category stored before, and sends this information to the Amazon app, "Please launch the *Activity* which is suitable to this information."

7) The Amazon app launches the *Activity* which is suitable for this information, creates a *Context* object, and associates this object with the *Activity*. Then it invokes the method *onCreate()* for the *Activity*.

---

* This process analysis is basically based on the Android 6.0 source code.

Up until now, we have seen the whole process from when the user clicks the Amazon App icon to the *Activity* of the Amazon app. From step 1 to 3, the communication is between the *Launcher* and the *AMS*, from step 4 to 7, the communication is between the Amazon app and the *AMS*.

This process involves a series of classes, as listed below. I will introduce them in the following sections:

- *Instrumentation*
- *ActivityThread*
- *H*
- *LoadedApk*
- *AMS*
- *ActivityManagerNative* (*AMN* for short) and *ActivityManagerProxy* (*AMP* for short)
- *ApplicationThread* (*APT* for short) and *ApplicationThreadProxy* (*ATP* for short)

2.5.2.1 Click the App Icon in Launcher *and Send a Message to the* AMS
In this section, we will introduce the process from when the user clicks on the app icon in the *Launcher* to when the *AMS* receives the message from the *Launcher*. There are six steps, shown in Figure 2.5.

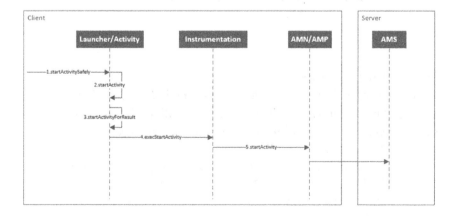

FIGURE 2.5   Process of *Launcher* informs the *AMS*.

Step 1 User clicks on the App icon in *Launcher*

The user clicks on the app icon in the *Launcher*; it invokes the method *startActivitySafely(Intent intent)* of *Activity*. The parameter is an *Intent* object; it carries the following information:

```
action = "android.intent.action.MAIN"
category = "android.intent.category.LAUNCHER"
cmp = "com.Amazon.activity.MainActivity"
```

Step 2 startActivity()

The method *startActivitySafely()* invokes the method *startActivity()* of *Activity*. There is a series of overload methods *startActivity()* defined in *Activity*. All these methods will invoke the method *startActivityForResult()*.

The method *startActivityForResult()* has two parameters. The first parameter is an *Intent* object; we pass an *Intent* object into this parameter with three pieces of information. The second parameter is the result of launching *Activity*. The *Launcher* doesn't care about whether the Amazon app launches successfully or not. So, the second parameter is set to –1 in this scenario. The code is as follows:

```
public void startActivity(Intent intent, @Nullable
Bundle options) {
    if (options != null) {
      startActivityForResult(intent, -1, options);
    } else {
      startActivityForResult(intent, -1);
    }
  }
```

Step 3 startActivityForResult()

Activity has a field *mInstrumentation*; its type is *Instrumentation*. In the method *startActivityForResult()* of *Activity*, it will invoke the method *execStartActivity()* of the object *Instrumentation*, shown as follows:

```
public void startActivityForResult(Intent intent, int
requestCode, @Nullable Bundle options) {
    // omit some codes
    Instrumentation.ActivityResult ar =
```

```
mInstrumentation.execStartActivity(
  this, mMainThread.getApplicationThread(),
  mToken, this,
  intent, requestCode, options);
}
```

In the code above, there is a variable named *mMainThread*, and its type is *ActivityThread*.

ActivityThread is the main thread and is also called the UI thread. It is created when the app starts and it represents the app.

Someone may mistakenly believe that the class *Application* is the core of the app. No, *Application* is a global class, although it's also created when the app is launched. *Application* is important in app development. It's open to the app developers, but *ActivityThread* is not open to the App developers. That's why app developers are familiar with *Application* but don't know *ActivityThread*.

ActivityThread has a secret. An Android app doesn't have the function *main()*. The function *main()* of each app is defined in *ActivityThread*.

Nearly all the applications have their own function *main()*, whether the application is written in Python, C#, C++, or Objective-C. It's the entry point of the application. But an Android app doesn't have the function *main()*. All the Apps share the same function *main()* defined in *ActivityThread*, shown as follows:

```
public final class ActivityThread {
  public static void main(String[] args) {
    Trace.traceBegin(Trace.TRACE_TAG_ACTIVITY_MANAGER,
    "ActivityThreadMain");
    SamplingProfilerIntegration.start();
    CloseGuard.setEnabled(false);
    Environment.initForCurrentUser();

    //omit some code
  }
}
```

An Android app doesn't have the function *main()*, so it's not a real application. It's only a package containing a lot of classes and resources.

Let's come back and focus on the parameters of the method *mInstrumentation.execStartActivity()*:

- *mMainThread.getApplicationThread()* is a *Binder* object and represents the process of *Launcher.*

- *mToken* is also a *Binder* object, represents the current *Activity* of *Launcher.*

These two parameters will be passed to the *AMS*. The *AMS* can then send the message back to the *Launcher* according to these two parameters.

Step 4 The Method *execStartActivity()* of Instrumentation
The method *execStartActivity()* of *Instrumentation* is used to launch a new *Activity*, shown as follows:

```
public class Instrumentation {

  public ActivityResult execStartActivity(
     Context who, IBinder contextThread, IBinder
     token, Activity target,
     Intent intent, int requestCode, Bundle options) {

   //omit some code

   try {
        // omit some code

     int result = ActivityManagerNative.getDefault()
       .startActivity(whoThread, who.
       getBasePackageName(), intent,
          intent.resolveTypeIfNeeded(who.
          getContentResolver()),
          token, target != null ? target.mEmbeddedID :
          null,
          requestCode, 0, null, options);

     // omit some code

   } catch (RemoteException e) {
        throw new RuntimeException("Failure from
        system", e);
   }
   return null;
  }
}
```

It invokes the method *ActivityManagerNative.getDefault().startActivity()*.

Step 5: The *AMN*'s getDefault Method
ActivityManagerNative (*AMN*) is widely used in the communication between the *AMS* and the four components, like *Activity*.

ServiceManager is a container. *AMN.getDefault()* obtains an object from the *ServiceManager* and then wraps this object into an *ActivityManagerProxy* object (*AMP*). The *AMP* is the proxy of the *AMS*.

The method *getDefault()* of the *AMN* returns an *IActivityManager* object. *IActivityManager* is an interface that implements *IInterface*, which defines all the lifecycle methods of the four components.

Both the *AMN* and the *AMP* implement the interface *IActivityManager*. The *AMS* inherits from *AMN*. Refer to Figure 2.6 for detail.

Step 6: startActivity() of the *AMP*
The *AMP* acts as the *Proxy* of *AIDL*, as introduced in Section 2.3. The method *startActivity()* of the *AMP* writes data to the *AMS* process and waits for the result from the *AMS*.

Then, the *Launcher* sends a message to the *AMS* and tells the *AMS* which app and which *Activity* to launch.

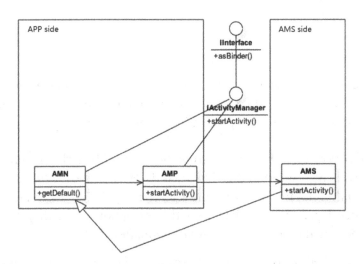

FIGURE 2.6   Class diagram of the *AMN/AMP*.

2.5.2.2 The AMS *Handles the Information from the* Launcher
First, let's have a look at the implementation of the method *startActivity()* of the *AMP*, shown as follows:

```
class ActivityManagerProxy implements IActivityManager
{
  public int startActivity(IApplicationThread caller,
  String callingPackage, Intent intent,
     String resolvedType, IBinder resultTo, String
     resultWho, int requestCode,
     int startFlags, ProfilerInfo profilerInfo, Bundle
     options) throws RemoteException {
    Parcel data = Parcel.obtain();
    Parcel reply = Parcel.obtain();
    data.writeInterfaceToken(IActivityManager.
    descriptor);
    data.writeStrongBinder(caller != null ? caller.
    asBinder() : null);
    data.writeString(callingPackage);

    mRemote.transact(START_ACTIVITY_TRANSACTION, data,
    reply, 0);
    reply.readException();
    int result = reply.readInt();
    reply.recycle();
    data.recycle();
    return result;
  }
}
```

The *AMS* is a system process. App developers have no permissions to modify code in the *AMS*. So, we don't spend much time in the *AMS*. Let's have a quick look what happens in the *AMS*.

1) *AMS* receives the message from the *AMN/AMP* in the app process; it will check whether this *Activity* is declared in the *AndroidManifest. xml*. If not found, it will throw an *ActivityNotFoundException*. App developers are familiar with this exception when they forget to declare the *Activity* in the *AndroidManifest.xml*.

2) The *AMS* sends a message to the *Launcher*, "Well, I got it. It's none of your business. Good Night." Now the *AMS* is *a Binder Client*, and the *Launcher* is the *Binder Server*.

When the *Launcher* sends a message to the *Anonymous Shared Memory* (*ASM*), the *Launcher* also send itself to the *AMS*. The *AMS* saves it as an *ActivityRecord* object.

ActivityRecord has an internal field; this field is an *ApplicationThreadProxy* object.

ApplicationThreadProxy (*ATP* for short) is a proxy; it's used to send data to the app process.

In the App process, the *ApplicationThread* (*APT* for short) is used to receive messages from the *AMS*.

Now the *AMS* can send a message back to the corresponding *Launcher*. Refer to Figure 2.7.

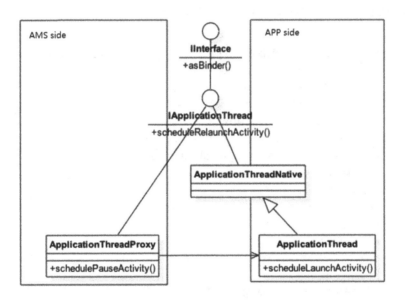

FIGURE 2.7   Class diagram of *IApplication* Thread.

2.5.2.3 The Launcher Goes to Sleep and Informs the AMS Again

After the *APT* receives a message from the *AMS*, it calls the method *sendThread()* of *ActivityThread* and sends a *PAUSE_ACTIVITY* message to the message queue of the *ActivityThread*, shown as follows (Figure 2.8). Let's focus on *ActivityThread*, as it's the main (UI) thread. Observe the following method *SendMessage()*:

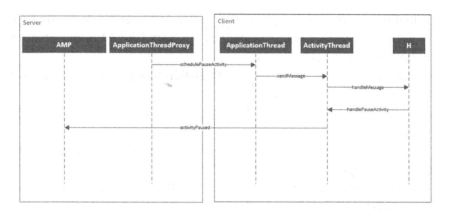

FIGURE 2.8 *Launcher* informs the *AMS* again.

```
private void sendMessage(int what, Object obj, int
arg1, int arg2, boolean async) {
  if (DEBUG_MESSAGES) Slog.v(
    TAG, "SCHEDULE " + what + " " +
    mH.codeToString(what)
    + ": " + arg1 + " / " + obj);
  Message msg = Message.obtain();
  msg.what = what;
  msg.obj = obj;
  msg.arg1 = arg1;
  msg.arg2 = arg2;
  if (async) {
    msg.setAsynchronous(true);
  }
  mH.sendMessage(msg);
}
```

In the code above, there is a variable *mH*. The type of *mH* is *H*. The class *H* inherits from the class *Handler*. *H* is used to dispatch messages to the corresponding method. The name of *H* is easy to remember.

H inherits from *Handler*, so it must implement the method *handleMessage()* of *Handler*. This method is a switch statement, which is used to handle different types of messages. *PAUSE_ACTIVITY* is one of these types.

All the messages sent to the four components will go through the method *handleMessage()* of *H*, shown as follows:

```
public final class ActivityThread {
  private class H extends Handler {

  //omit some code
  public void handleMessage(Message msg) {
      if (DEBUG_MESSAGES)
      Slog.v(TAG, ">>> handling: " + codeToString(msg.
      what));

    switch (msg.what) {
    case PAUSE_ACTIVITY:
      Trace.traceBegin(Trace.TRACE_TAG_ACTIVITY_
      MANAGER, "activityPause");
      handlePauseActivity((IBinder)msg.obj, false,
      (msg.arg1&1) != 0, msg.arg2, (msg.arg1&2) != 0);
      maybeSnapshot();
      Trace.traceEnd(Trace.TRACE_TAG_ACTIVITY_MANAGER);
          break;
    };
    // omit some code
  }
}
```

When the method *handleMessage()* of *H* handles the *PAUSE_ACTIVITY* message, it will invoke the method *handlePauseActivity()* of the *ActivityThread*. This method executes the following logic:

- Get the current *Activity* of the *Launcher* and execute its method *onPause()*. In the *ActivityThread* there is a collection named *mActivities*; this collection stores all the open *Activities*. The current *Activity* is in this collection.

- Inform the *AMS*, "I have fallen asleep."

2.5.2.4 The AMS Creates a New Process

The second time the *AMS* receives a message from the *Launcher* it will try to open the Amazon app.

If the Amazon app is already open, but in the background, the *AMS* will notify the Amazon app to wake up and become visible to the user.

If the Amazon app is not open, the *AMS* needs to create a new process. It invokes the method *Process.start()* to create a new process.

After the new process is created, the *AMS* will invoke the function *main()* to launch the Amazon app, which is the entry point to the app, shown as follows:

```
int pid = Process.start("android.app.ActivityThread",
    mSimpleProcessManagement ? app.processName : gid,
    debugFlags, null);
```

2.5.2.5 Start a New Process and Inform the AMS

When the *AMS* creates a new process, it will create an *ActivityThread* for this process (Figure 2.9).

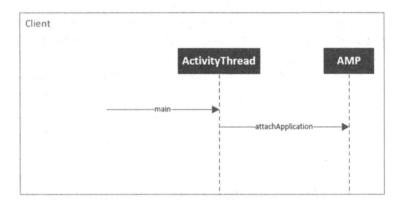

FIGURE 2.9   Invoke the function *main()* and inform the *AMS*.

After the *AMS* creates the UI thread, it will execute the function *main()* of the *ActivityThread*, including two logics:

1) Create *MainLooper*. Loopers are used to handle messages in the Android system. *MainLooper* is a special looper. Each app has only one *MainLooper*, which is an endless loop to receive messages from the system.

2) Create an *Application*. Application is created during this step. That is why the methods *onCreate()* and *attach()* are executed earlier than the other methords of *Activity*, and why we usually initialize global variables in these two methods.

After receiving the *BIND_APPLICATION* message, the UI thread creates a *LoadedApk* object (which stores the information for the current *apk*), then creates a *ContextImpl* object, and then uses reflection grammar to create *Application* and invoke the method *attach()* of *Application* and finally invokes the method *onCreate()* of Application to initialize some variables.

Lastly, the new process informs the *AMS*, "I'm ready," and sends itself as a parameter to the *AMS*.

2.5.2.6 The AMS *Tells the New App Which* Activity *to Launch*

AMS receives the token from the new process; in this scenario, the new process is used for the Amazon app.

During the previous steps, the *AMS* stores the information about which app and which *Activity* is to be launched. Now it's time to use this information.

The *AMS* fetches this information and sends it to the new process.

2.5.2.7 The Amazon App Starts an Activity

This is the final step of the whole process (refer to Figure 2.10).

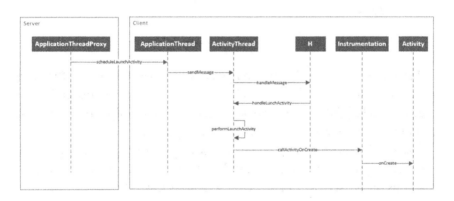

FIGURE 2.10   Launch a new *Activity* in the Amazon app.

The new process, in this scenario the Amazon app, receives a message from the *AMS* by *ApplicationThread* and handles this message with the method *handleMessage()* of *H*; the logic is in the switch statement. In this scenario, the type of message is *LAUNCH_ACTIVITY*, shown as follows:

```
public final class ActivityThread {
    private class H extends Handler {

    //omit some code
    public void handleMessage(Message msg) {
```

```
if (DEBUG_MESSAGES) Slog.v(TAG, ">>> handling: " +
codeToString(msg.what));

switch (msg.what) {
    case LAUNCH_ACTIVITY: {
        Trace.traceBegin(Trace.TRACE_TAG_ACTIVITY_
        MANAGER, "activityStart");
        final ActivityClientRecord r =
        (ActivityClientRecord) msg.obj;

        r.packageInfo = getPackageInfoNoCheck(
        r.activityInfo.applicationInfo,
        r.compatInfo);
        handleLaunchActivity(r, null);
        Trace.traceEnd(Trace.
        TRACE_TAG_ACTIVITY_MANAGER);
    } break;
    // omit some code
}
}
```

Let's have a look at the method *getPackageInfoNoCheck()* in the code above. This method is used to fetch all the resources of an *apk* and put it into an *r.packageInfo*. The type of *packageInfo* is *LoadedApk*. In the plug-in technique, we can replace it with our own object.

ActivityThread doesn't handle the message directly. It sends all the messages to *H*. *H* is responsible for dispatching messages to the different methods of the *ActivityThread*. Refer to Figure 2.11.

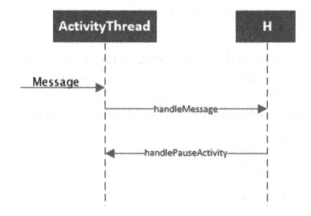

FIGURE 2.11   Interaction between *ActivityThread* and *H*.

Now let's have a look at the logic in the method *handleLaunchActivity()*:

1) Invoke the method *newActivity()* of *Instrumentation* to create an instance of the *Activity* to be launched.

2) Create a *Context* object and associate it with this *Activity*.

3) Invoke the method *callActivityOnCreate()* of *Instrumentation* to execute the method *onCreate()* of this *Activity*.

Up until now, the home page of the Amazon app has been accessed by the user. There are many handshakes in this process. The *Launcher* and the Amazon app communicates frequently with the *AMS* based on *Binder*.

2.6 NAVIGATION IN APP

Navigation from *ActivityA* to *ActivityB* is the same as the mechanism I introduced in Section 2.5.

If *ActivityA* and *ActivityB* are running in the same process, the whole process will be simpler, because the *AMS* doesn't need to create a new process for *ActivityB*. There is a total of five steps in this process, as follows:

1) *ActivityA* sends a message to the *AMS*, "I want to launch *ActivityB*."

2) *AMS* saves the information of *ActivityB*, and then informs the app, "Got it."

3) *ActivityA* invokes its method *onPause()*, and informs the *AMS*, "I'm sleeping."

4) The *AMS* finds that *ActivityA* and *ActivityB* are in the same process, and the *AMS* notifies the app to launch *ActivityB* directly.

5) The app launches *ActivityB*.

Refer to Figure 2.12 for details.

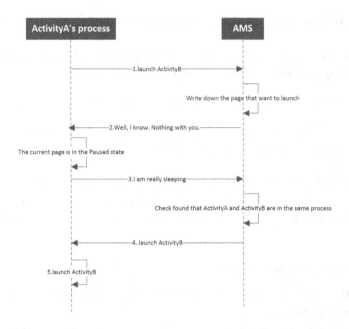

FIGURE 2.12 Launch a new *Activity* in the same process.

2.7 FAMILY OF CONTEXT

Activity, *Service*, and *Application* have the same "ancestor;" they are a family, as shown in Figure 2.13.

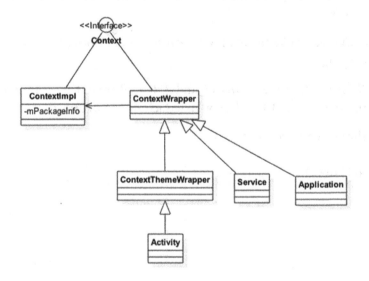

FIGURE 2.13 *Context* Family.

Because *Activity* needs *Theme*, *Activity* inherits from the *ContextThemeWrapper* directly, *ContextThemeWrapper* inherits from *ContextWrapper*. *Service* and *Application* also inherit from *ContextWrapper*.

ContextWrapper is a wrapper class without logic. The real logic is inside *ContextImpl*.

The number of *Contexts* in an app is equal to the number of *Services* + number of *Activities* + 1 (*Application* itself).

Take *Activity* as an example to have a look at the relationships and differences between *Activity* and *Context*.

For example, to jump from one *Activity* to another *Activity*, the code is shown as follows:

```
btnNormal.setOnClickListener(new View.
OnClickListener() {
    @Override
    public void onClick(View view) {
        Intent intent = new Intent(Intent.Action.VIEW);
        intent.addFlags(Intent.FLAG_ACTIVITY_NEW_TASK);
        intent.setData(Uri.parse("https://www.baidu.com"));
        startActivity(intent);
    }
});
```

Another way of jumping from one *Activity* to another *Activity* is to invoke the method *getApplicationContext()* to get a *Context* object, and then invoke the method *startActivity()* of this *Context* object, shown as follows:

```
btnNormal.setOnClickListener(new View.
OnClickListener() {
    @Override
    public void onClick(View view) {
        Intent intent = new Intent(Intent.Action.VIEW);
        intent.addFlags(Intent.FLAG_ACTIVITY_NEW_TASK);
        intent.setData(Uri.parse("https://www.baidu.com"));
        getApplicationContext().startActivity(intent);
    }
});
```

The differences between these two methods are shown in Figure 2.14.

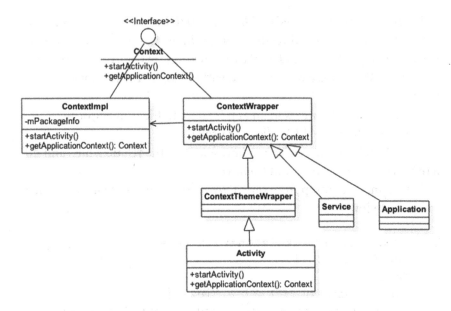

FIGURE 2.14   Two kinds of *startActivity()*.

The method *startActivity()* of *Context*, invokes the method *execStartActivity()* of *mInstrumentation*, *mInstrumentation* is a field of *ActivityThread*.

We invoke the method *getApplicationContext()* to get a *Context* Object; its method *startActivity()* is defined in *ContextImpl*, and it invokes the method *execStartActivity()* of *mInstrumentation* of *ActivityThread* too, shown as follows:

```
class ContextImpl extends Context {
    @Override
    public void startActivity(Intent intent, Bundle
    options) {
        warnIfCallingFromSystemProcess();
        if ((intent.getFlags()&Intent.FLAG_ACTIVITY_
        NEW_TASK) == 0) {
            throw new AndroidRuntimeException(
                "Calling startActivity() from outside of
                an Activity "
                + " context requires the FLAG_ACTIVITY_
                NEW_TASK flag."
                + " Is this really what you want?");
        }
        mMainThread.getInstrumentation().
        execStartActivity(
```

```
            getOuterContext(), mMainThread.
            getApplicationThread(), null,
            (Activity) null, intent, -1, options);
    }
}
```

2.8 SERVICE

Service has two processes, one is a launching process, and the other is a binding process. All app developers are familiar with these two mechanisms, as shown in Figure 2.15.

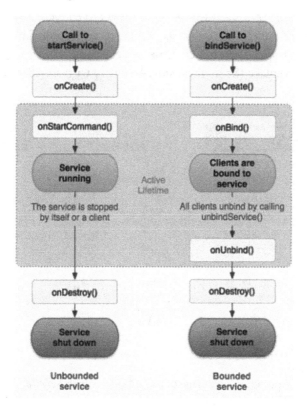

FIGURE 2.15   Launching and binding process of *Service.*

2.8.1 Start *Service* in a New Process

Let's have a look at the *Service* launching process. Suppose the *Service* to be launched is in a new process, it's divided into five steps:

1) The app sends a message to the *AMS*, "I want to launch a *Service* in a new process."

2) The *AMS* checks whether this *Service* is already launched. If no, the *AMS* saves the information of this *Service* and then *creates* a new process to launch this *Service*.

3) The new process informs the *AMS*, "I am ready."

4) The *AMS* notifies the new process to launch this *Service*.

5) The new process launches this *Service*.

Let's begin step by step.

2.8.1.1 The App Sends a Message to the AMS to Launch Service

The process of launching *Service* is similar to an *Activity* launching process; we won't spend too much time to repeating the same process. In Figure 2.16, I replace the method *startActivity()* with *startService()*.

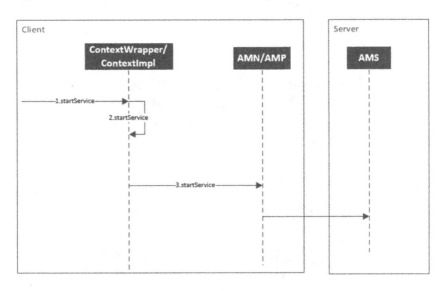

FIGURE 2.16 App sends a message to the *AMS* to launch *Service*.

2.8.1.2 The AMS Creates a New Process

When the *AMS* receives the message from the app process (*AMN/AMP*), the *AMS* will check whether this *Service* is declared in the *AndroidManifest. xml*. If no, the *AMS* will throw an exception to the app process; otherwise, the *AMS* will create a new process to launch this *Service*.

In the *AMS*, each *Service* is saved as a *ServiceRecord* object.

2.8.1.3 Start a New Process and Inform the AMS

After the new process is created, it will create a new *ActivityThread* object, and then pass this object to the *AMS* by *AMN/AMP*. It informs the *AMS*, "I'm ready. Can I help you?"

2.8.1.4 The AMS Sends Information to the New Process

When the *AMS* receives the *ActivityThread* object, the *AMS* will convert this object into an *ApplicationThreadProxy*(*ATP* for short) object.

During the previous steps, the *AMS* stores the information about which *Service* is to be launched. Now it's time to use this information.

The *AMS* fetches this information and sends it to the new process.

2.8.1.5 New Process to Launch Service

The new process receives the *AMS* information by *ApplicationThread*. With the help of *ActivityThread* and *H*, the method *onCreate()* of this *Service* will be executed (Figure 2.17).

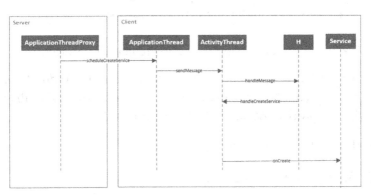

FIGURE 2.17   Launch a new *Service*.

The new process will create a *Context* object and associate it with this *Service*.

Let's focus on the method *handleCreateService()* of *ActivityThread*, shown as follows:

```
private void handleCreateService(CreateServiceData
data) {
  LoadedApk packageInfo = getPackageInfoNoCheck(
      data.info.applicationInfo, data.compatInfo);
  Service = null;
  try {
```

```
    java.lang.ClassLoader cl = packageInfo.
    getClassLoader();
    service = (Service) cl.loadClass(data.info.name).
    newInstance();
  }

  //omit some code
}
```

We will find that these codes are similar to the method *handleLaunchActivity()* introduced earlier. First, fetch package information using the *PMS*, then create a *LoadedApk* object and get its own *ClassLoader*, finally use this *ClassLoader* to load a *Service*.

Now we have launched a *Service* in a new process.

2.8.2 Start a *Service* in the Same Process

If we want to launch a *Service* in the current process, the launching process will be simplified in the following three steps:

1) The app sends a message to the *AMS* to launch a *Service*.

2) The *AMS* checks if this *Service* is declared in the *AndroidManifest. xml*. If no, it will throw an exception; otherwise, the *AMS* will inform the app process to launch the *Service* directly.

3) The app launches the *Service*.

2.8.3 Bind a *Service* in the Same Process

Let's have a look at the *Service* binding process in the current process. The process has the following five steps:

1) The app sends a message to the *AMS*, "I want to bind a *Service*."

2) The *AMS* checks if this *Service* is declared in the *AndroidManifest. xml*. If yes, the *AMS* will inform the app process to launch this *Service*, and then bind this *Service* later. The *AMS* will send two messages to the app.

3) The app receives the first message from the *AMS* and launches this *Service*.

4) The app receives the second message from the *AMS*; it will bind this *Service* and send a *Binder* object to the *AMS*.

5) When the *AMS* receives the *Binder* object from the app process, it will send the *Binder* object back to the app process.

Although we want to launch and bind a *Service* in the current app process, the current app process still needs to communicate with the *AMS* frequently.

Let's study this process step by step.

2.8.3.1 The App Sends a Message to the AMS to Bind a Service
The process in Figure 2.18 is the same as Section 2.8.1.1, so we won't spend much time on it. I simply replace the method *startService()* with *bindService()*.

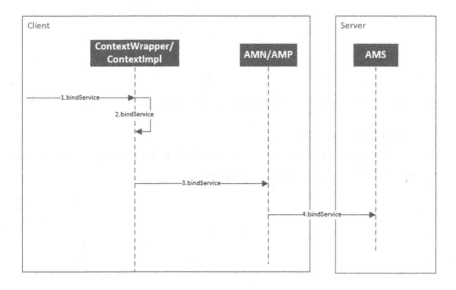

FIGURE 2.18   App sends a message to the *AMS*.

We pass the *ServiceConnection* object as a parameter into the method *bind()* of the *Service*. If the binding is successful, the method *onService-Connected()* of this *ServiceConnection* object will be invoked.

2.8.3.2 The AMS Sends Two Messages to the App Process
The process is the same as in Section 2.8.1.2.

2.8.3.3 The App Receives the First Message
The process is the same as Section 2.8.1.5.

2.8.3.4 The App Receives the Second Message and Sends a Binder Object to the AMS

The *AMS* will send two messages to the app process; the second message is used to bind the *Service*. The method *bind()* of *Service* returns a *Binder* object. The app client will send this *Binder* object to the *AMS* (Figure 2.19).

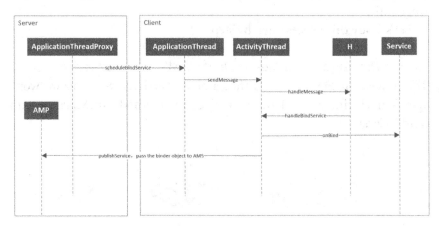

FIGURE 2.19   Handle the second message in the app process.

2.8.3.5 AMS Informs the App

The *AMS* informs the app process to establish if the binding is successful. It's also based on *AIDL*, shown in Figure 2.20.

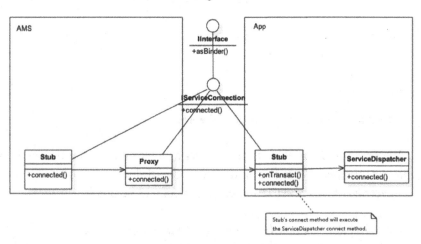

FIGURE 2.20   The *AMS* informs the app to establish the binding is successful.

As shown in Section 2.8.1, when the binding is successful, the method *onServiceConnected()* of this *ServiceConnection* object will be invoked.

For the last step in Figure 2.20, the method *connect()* of *ServiceDispatcher* is invoked. It will invoke the method *onServiceConnected()* of *ServiceConnection*.

2.9 BROADCASTRECEIVER

BroadcastReceiver is a broadcaster; it's also called *Receiver* for short.

Many app developers say that they haven't used a *Receiver*. In fact, *Receiver* and *Service* are widely used in music player apps. When the user clicks on the play button, it will notify the background process to play music.

On the other hand, when music plays to the end of a song, and another song begins to play, the background process will send a message to the app process, that's why the app switches the description of the song from one to another.

So, the principle of the music app is the communication between *Activity* and *Service* in different processes, shown in Figure 2.21.

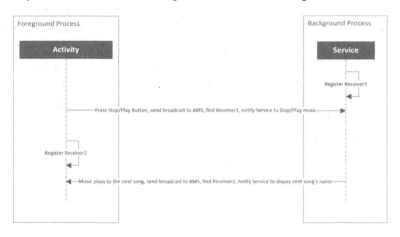

FIGURE 2.21 Two *Receivers* for a music player.

The *Receiver* has two types, one is a *Static Receiver*, and the other one is a *Dynamic Receiver*.

Static Receiver must be declared in the *AndroidManifest.xml*, shown as follows:

```
<receiver android:name=".Receiver1">
    <intent-filter>
        <action android:name="baobao" />
    </intent-filter>
</receiver>
```

We can also register a *Dynamic Receiver* in code, shown as follows:

```
Receiver2 receiver2 = new Receiver2();
IntentFilter filter = new IntentFilter();
filter.addAction("UpdateActivity");
registerReceiver(receiver2, filter);
```

We can send a broadcast, shown as follows:

```
Intent intent = new Intent("UpdateActivity");
sendBroadcast(intent);
```

These two types have the same function. So, we can change all the *Static Receivers* to *Dynamic Receivers*.

But there is a small difference between these two types. We can send a broadcast to a *Static Receiver* even if the app is not launched. But this feature is not suitable for a *Dynamic Receiver.*

Now we have a look at the process where the *Receiver* communicates with the *AMS*. It consists of two parts. One is the registration; the other is a broadcast.

Let's take the music player app as an example. We register *Receiver* in *Activity* and send a broadcast in *Service*. When *Service* plays the next song, it will notify *Activity* to modify the name of the current song.

2.9.1 Registration

Figure 2.22 shows the process of registration.

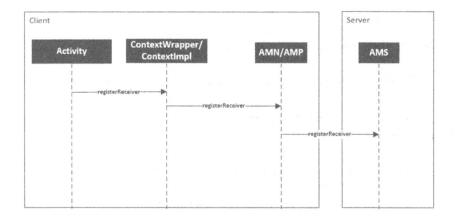

FIGURE 2.22   Process of register *Receiver.*

1) Register the *Receiver* and notify the *AMS*.

In *Activity*, we can invoke the method *registerReceiver()* of *Context*; it will pass the *Receiver* and its *IntentFilter* onto the *AMS* via *AMN/AMP*.

When we create a *Receiver*, we need to specify a *IntentFilter* for this *Receiver*. *IntentFilter* is the character of *Receiver*; it is used to find which *Receiver* (one or more) is suitable to the broadcast.

In the method *registerReceiver()* of *Context*, it uses *PMS* to get the information of the *apk*, which is a *LoadedApk* object. The method *getReceiver-Dispatcher()* of this *LoadedApk* object, encapsulates *Receiver* as a *Binder* object which implements the interface *IIntentReceiver*.

2) After the *AMS* receives the message including the *Receiver* and *IntentFilter*, the *AMS* will store this information in a list. This list contains all the *Dynamic Receivers*.

But when does the *Static Receiver* register with the *AMS*? The answer is upon app installation. During the installation of the app, *PMS* will parse the four components declared in the *AndroidManifest.xml*, including *Static Receiver*. *PMS* will save all the *Static Receivers* in a list.

Dynamic Receiver and *Static Receiver* exist in the different lists. When someone sends a broadcast, these two lists will be merged together into a new list. The *ASM* iterates this new list to find which *Receiver* is suitable for the broadcast.

2.9.2 Send a Broadcast

The process of sending a broadcast is as follows:

1) In *Service* or *Activity*, send a broadcast with an *IntentFilter* to the *AMS* via the *AMM/AMP*, as shown in Figure 2.23.

2) When the *AMS* receives this broadcast, the *AMS* will find the corresponding *Receivers* in the *Receiver* list. The *AMS* puts all the *Receivers* into the broadcast queue and sends a message to the message queue.

When the Android system handles this message in the message queue, the *AMS* will find the suitable *Receivers* in the broadcast queue and send a broadcast to these *Receivers*.

Why do we need two queues? Because the sending messages and the receiving messages are asynchronous. Each time the *AMS* receives a

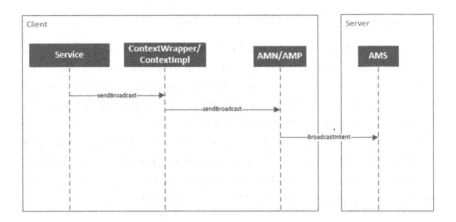

FIGURE 2.23 Service sends a broadcast to the *AMS*.

broadcast, it will throw the broadcast into a broadcast queue. But the *AMS* doesn't care about whether the message is sent successfully or not.

The *AMS* sends a broadcast to the app. It's also based on *AIDL*.

3) When the app receives the broadcast from the *AMS*, it doesn't send the broadcast to the *Receiver* directly. The app encapsulates the broadcast as a message and sends this message to the message queue of the *ActivityThread*. When this message is handled in the message queue, it will send a broadcast to the corresponding *Receiver*, as shown in Figure 2.24.

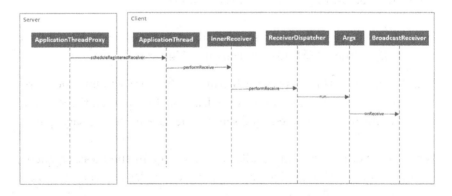

FIGURE 2.24 App handles a broadcast.

2.10 CONTENTPROVIDER

ContentProvider is also called *CP* for short. App developers are not familiar with *ContentProvider*.

ContentProvider is used to transfer large data (size > 1M). So *ContentProvider* is widely used in Android ROM development to transfer large data from one process to another. But in an app, large data is stored in the remote server; we can invoke the remote API to fetch this data rather than *ContentProvider*.

There is an exception to everything. Sometimes the app will read the data of the Address Book or SMS. At this moment we need to use *ContentProvider*. The data of the Address Book or SMS is provided in the form of *ContentProvider*.

Let's have a quick review of how to use *ContentProvider* in an app, as shown in Figure 2.25.

We need to write code in both app1 and app2, as follows:

1) Define *ContentProvider* in app1

Define a *ContentProvider* in app1: give it the name *MyContentProvider*, and declare *MyContentProvider* in the *AndroidManifest.xml*. We need to

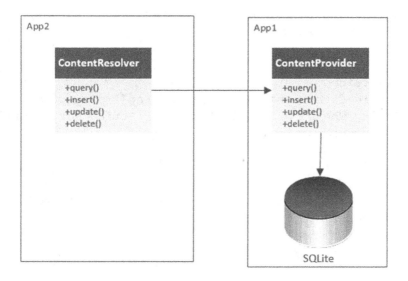

FIGURE 2.25   App2 accesses the *ContentProvider* provided by app1.

implement four methods, *insert()*, *update()*, *delete()*, and *query()*, for the *ContentProvider*, shown as follows:

```
<provider
  android:name=".MyContentProvider"
     android:authorities="baobao"
     android:enabled="true"
     android:exported="true"></provider>

public class MyContentProvider extends ContentProvider {
  public MyContentProvider() {
  }

  @Override
  public boolean onCreate() {
    //Omit some code
  }

  @Override
  public String getType(Uri uri) {
    //Omit some code
  }

  @Override
  public Uri insert(Uri uri, ContentValues values) {
    // Omit some code
  }

  @Override
  public Cursor query(Uri uri, String[] projection,
  String where, String[] whereArgs, String sortOrder){
    // Omit some code
  }

  @Override
  public int delete(Uri uri, String where, String[]
  whereArgs) {
    // Omit some code
  }
```

```
@Override
public int update(Uri uri, ContentValues values,
String where, String[] whereArgs){
  // Omit some code
  }
}
```

2) Use *ContentProvider* in App2

In app2, we access the *ContentProvider* defined in app1 using *ContentResolver*; it also provides four methods: *insert()*, *update()*, *delete()*, and *query()*, which is used to access the *ContentProvider* defined in app1:

```
public class MainActivity extends Activity {
  ContentResolver contentResolver;
  URI uri;

  @Override
  protected void onCreate(Bundle savedInstanceState) {
    super.onCreate(savedInstanceState);
    setContentView(R.layout.activity_main);

    uri = Uri.parse("content://baobao/");
    contentResolver = getContentResolver();
  }

  public void delete(View source) {
    int count = contentResolver.delete(uri, "delete_
    where", null);
    Toast.makeText(this, "delete uri:" + count, Toast.
    LENGTH_LONG).show();
  }

  public void insert(View source) {
    ContentValues values = new ContentValues();
    values.put("name", "jianqiang");
    Uri newUri = contentResolver.insert(uri, values);
    Toast.makeText(this, "insert uri:" + newUri,
    Toast.LENGTH_LONG).show();
  }

  public void update(View source) {
    ContentValues values = new ContentValues();
```

```
    values.put("name", "jianqiang2");
    int count = contentResolver.update(uri, values,
    "update_where", null);
    Toast.makeText(this, "update count:" + count,
    Toast.LENGTH_LONG).show();
  }
}
```

How to debug *ContentProvider*? Run app1 and app2 both in debug mode, and we can debug from app2 to app1.

Each *ContentResolver* has its own *URI*, which is declared in the *AndroidManifest.xml*, shown as follows:

```
<provider
  android:name=".MyContentProvider"
    android:authorities="baobao"
    android:enabled="true"
    android:exported="true"></provider>
```

URI is the identity of the *ContentProvider*; it's unique. When we want to invoke the *CRUD* methods of this *ContentProvider*, we need to specify *URI* as follows:

```
uri = Uri.parse("content://baobao/");
```

In the next section, I will talk about the communication mechanism between the *CRUD* methods of *ContentResolver* and the *AMS*.

2.10.1 The Essence of the *ContentProvider*

ContentProvider is the engine of the SQLite database.

Different data sources have different data formats and different API, such as SMS and the Address Book. But users want to view this data in the same format. So we write a *ContentProvider* to encapsulate the different data sources. *ContentProvider* always provides four methods, *insert()*, *update()*, *delete()*, and *query()*, also called *CRUD* for short.

2.10.2 The *ASM*

ContentProvider reads data using *Anonymous Shared Memory* (*ASM* or *Ashmem* for short), refer to Figure 2.26 for details. The *Server* provides data; the *Client* uses data.

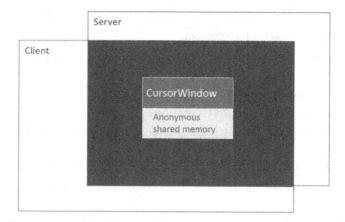

FIGURE 2.26   Structure of the *ASM*.

Figure 2.27 shows communication between the *Client* and the *Server*.

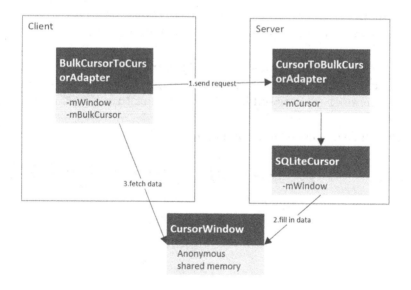

FIGURE 2.27   Class diagram of the *ASM*.

The process is as follows:

1) There is a *CursorWindow* object inside the *Client*. The *Client* sends the request to the *Server*; the request carries a *CursorWindow* object. Now this object is empty.

2) The *Server* receives the request, collects the data, and puts the data into the *CursorWindow* object.

3) The *Client* reads its internal *CursorWindow* object and gets the data.

Thus, this *CursorWindow* object is the *ASM*.

For example, each house has a mailbox. The postman throws mail into the mailbox; we fetch mail in the mailbox. The mailbox is the *ASM*.

2.10.3 Communication between *ContentProvider* and the *AMS*

Now let's have a look at the communication between a *ContentProvider* and the *AMS*.

Here I use some pictures rather than thousands of lines of code to make this section more comprehensible.

For example, app2 wants to use the method *insert()* of a *ContentProvider* defined in app1, shown as follows:

```
ContentResolver = getContentResolver ();
URI uri = Uri.parse("content://baobao/");

ContentValues values = new ContentValues();
values.put("name", "jianqiang");
URI newUri = contentResolver.insert(uri, values);
```

The whole process is shown in Figure 2.28.

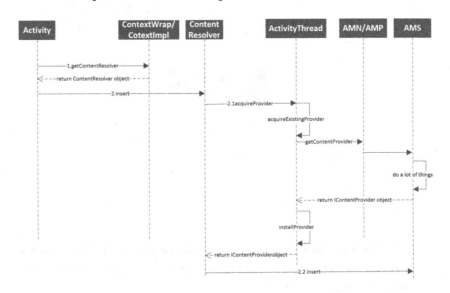

FIGURE 2.28   Communication between *ContentProvider* and the *AMS*.

Let's analyze this process step by step:

1) App2 sends a message to the *AMS*, "I want to access *ContentProvider* in app1."

2) The *AMS* checks whether the *ContentProvider* in app1 has been started. If not, it will start a new process to launch app1, start a *ContentProvider*, wrap this *ContentProvider* to a proxy object, and finally return a proxy object to app2.

3) App2 gets the proxy object of a *ContentProvider*, whose type is *IContentProvider*. The app then invokes the *CRUD* methods of this proxy object and will transfer data or modify data from the *ASM*.

Up to now, I have introduced the underlying knowledge of *ContentProvider*.

2.11 THE *PMS* AND APP INSTALLATION PROCESS

2.11.1 PMS Introduction

PMS (PackageManagerService) is used to obtain the information of an *apk*.

In the previous sections, we analyzed the communication between the four components and the *AMS*; as we have shown, the *AMS* uses the *PMS* to load information from an *apk*, and encapsulate this information into a *LoadedApk* object, and then we can fetch all the components declared in the *AndroidManifest.xml*.

During the process of app installation, the *apk* will be stored in the folder *Data/App*.

I always wondered why the app is not unzipped during app installation. It's easy to find any resource in the resource folder. In fact, reading resources from *apk* directly is quicker. We can find the implementation of this logic in the Android system. It's written in C++.

Every time an app reads resources from the *apk*; it will parse *resources.arsc* in the *apk.resources.arsc* stores all the information of resources, including the address, size, and other properties of the resources. It also stores the mapping between resource ID and the real name of the resource. Which means the app can fetch the resource using the resource ID quickly.

2.11.2 App Installation Process

Android system uses the *PMS* to parse the *AndroidManifest.xml* in this *apk*, including the following content:

- Information of all the components, such as *Static Receivers*.

- Assign *userId* and *userGroupId* to this app.

- At the end of the app installation process, the above information is stored in an *XML* file. This file will be reused for the next installation.

The Android system has an interesting feature in that the *PMS* will install all the apps again when the Android system reboots. There are four steps to this, as shown in Figure 2.29.

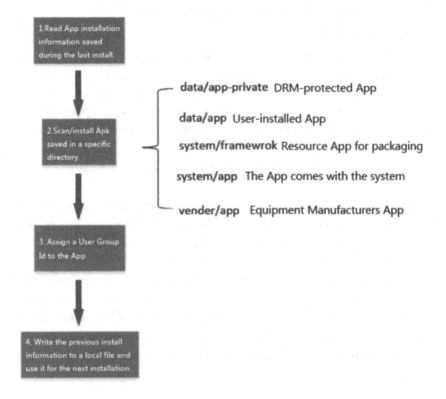

FIGURE 2.29 App installation process.

2.11.3 *PackageParser*

After the Android system reboots, all the apps in the Android system will be reinstalled by the *PMS* again.

The first time the app is installed in the Android system, this work is also completed by the *PMS*.

The *PMS* is the system process in the Android system.

Now I will introduce an important class, *PackageParser*. It's an important class in this book because the *PMS* uses *PackageParser* to parse the information in the *AndroidManifest.xml*.

The method *parsePackage()* of *PackageParser* has a parameter *apkFile*. We can use the current *apk* file or an external *apk* file as the parameter for this method.

Unfortunately, *PackageParser* is not open to app developers; we must use reflection to invoke the method *parsePackage()* of *PackageParser* to get the components declared in the *AndroidManifest.xml*.

The return value of *parsePackage()* is a *Package* object, which stores information of the four components in the *AndroidManifest.xml*. This class is of no use to us. So we always convert it to a *PackageInfo* object using the method *generatePackageInfo()* of *PackageParser*.

Now let's have a look at the relationships in the *PMS* family, as shown in Figure 2.30.

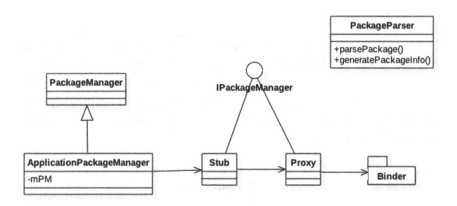

FIGURE 2.30 Class diagram of *PackageParser*.

2.11.4 *ActivityThread* and *PackageManager*

All roads lead to Rome. App developers also use the method *getPackage-Manager()* of *Context* to get information about the current app.

The concrete logic of the method *getPackageManager()* of *ContextImpl*, shown as follows:

```
class ContextImpl extends Context {
    private PackageManager mPackageManager;

    @Override
    public PackageManager getPackageManager() {
      if (mPackageManager != null) {
         return mPackageManager;
      }

      IPackageManager pm = ActivityThread.
      getPackageManager();
      if (pm != null) {
         // Doesn't matter if we make more than one
         instance.
         return (mPackageManager = new
         ApplicationPackageManager(this, pm));
      }

      return null;
    }
}
```

The method *getPackageManager()* returns an *ApplicationPackage Manager* object, which is a subclass of *PackageManager*.

ApplicationPackageManager is the decorator of the object *pm*; *pm* is the return value of the method *getPackageManager()* of the *ActivityThread*. The real logic is in the method *getPackageManager()* of the *ActivityThread*, shown as follows:

```
public final class ActivityThread {
private static ActivityThread sCurrentActivityThread;

    static IPackageManager sPackageManager;

    public static IPackageManager getPackageManager() {
      if (sPackageManager != null) {
```

```
        return sPackageManager;
    }
    IBinder b = ServiceManager.getService("package");
    sPackageManager = IPackageManager.Stub.
    asInterface(b);
    return sPackageManager;
  }
}
```

In Section 2.2, we introduced the *ServiceManager*, which is a dictionary storing various system services. For example, *Clipboard* is a system service stored in *ServiceManager*, its key is "clipboard"; the *PMS* is also a system service in *ServiceManager*, its key is "package."

IPackageManager is an *AIDL* file. Refer to Figure 2.30 for details.

ApplicationPackageManager doesn't communicate with *Binder* directly. It has a field, *mPM*, and the type of *mPM* is *IPackageManager*. Refer to the concept of AIDL (Section 2.3). *mPM* is the *Stub* of the *IPackageManager*.

According to the introduction in Section 2.3, we find that the following statements return the same object:

- The method *getPackageManager()* of *Context*.

- The method *getPackageManager()* of *ActivityThread*.

- The field *sPackageManager* of *ActivityThread*.

- The field *mPM* of *ApplicationPackageManager*.

All these statements are the proxy object of the *PMS* in the app process. We can use these statements to retrieve information about the current app, especially the information of the four components.

2.12 CLASSLOADER

It's time to introduce the family of *ClassLoader*. The class *ClassLoader* is the ancestor of this family, as shown in Figure 2.31.

Let's focus on *PathClassLoader*, *DexClassLoader*, and their parent class, *BaseDexClassLoader*.

DexClassLoader is a simple class, and it has only one constructor, shown as follows.

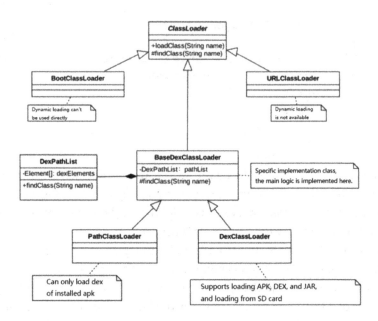

FIGURE 2.31   Family of *ClassLoader*.

```
public class DexClassLoader extends BaseDexClassLoader {
    public DexClassLoader(String dexPath, String
    optimizedDirectory,
        String librarySearchPath, ClassLoader parent) {
        super(dexPath, new File(optimizedDirectory),
        librarySearchPath, parent);
    }
}
```

In this constructor, it invokes its parent constructor directly. The second parameter of the constructor is the path of the *dex/apk*.

PathClassLoader is also a simple class, and it has only two constructors, shown as follows.

```
public class PathClassLoader extends BaseDexClassLoader {
    public PathClassLoader(String dexPath, ClassLoader
    parent) {
      super(dexPath, null, null, parent);
    }

    public PathClassLoader(String dexPath, String
    librarySearchPath, ClassLoader parent) {
      super(dexPath, null, librarySearchPath, parent);
    }
}
```

This constructor invokes its parent constructor directly. It passes *null* as the second parameter.

So we should use *DexClassLoader* to load a *dex/apk* file, but *PathClassLoader* is only used in the Android system.

2.13 PARENT-DELEGATION

Parent-Delegation is based on the family *ClassLoader.*

Let's have a look at the class *ClassLoader,* which is the ancestor in this family. The constructor of *ClassLoader* is as follows:

```
ClassLoader(ClassLoader parentLoader, boolean
nullAllowed) {
  if (parentLoader == null && !nullAllowed) {
    throw new NullPointerException("parentLoader ==
    null && !nullAllowed");
  }
  parent = parentLoader;
}
```

The first parameter of the constructor is still a *ClassLoader* object. This parameter will be passed to the internal field named *Parent* of the current *ClassLoader.*

When *DexClassLoader* loads a class, firstly, it will delegate its parent *BaseDexClassLoader* to load this class, and then *BaseDexClassLoader* will delegate its parent *ClassLoader* to load this class. If *BaseDexClassLoader* and *ClassLoader* can't load this class, *DexClassLoader* will load this class by itself.

It will improve the performance of loading a class. If the parent *ClassLoader* has loaded the class before, the Android system will use the parent *ClassLoader* to load this class directly. *Parent-Delegation* is widely used in plug-in frameworks. We use *DexClassLoader* to load the class of plug-ins.

2.14 MULTIDEX

If the version of the Android system is less than 5.0, app developers always meet the following issues:

```
Conversion to Dalvik format failed: Unable to execute
dex: method ID not in [0, 0xffff]: 65536
```

This issue is also called 65536. It occurs when the count of methods in an app is more than 65536.

We can write so many methods into an app, but we always use the third party SDK and compile it into the app. These SDKs have a lot of features with tens of thousands of methods, but we only use one or two methods in our app. That's why the count of methods exceeds 65536.

We can use ProGuard to reduce unused methods. The number of methods in an app will be less than 65,536, but we only use ProGuard in release mode. We must face this issue in debug mode.

The root cause of 65536 is in the older version of Android systems, such as 4.4, and there is a variable to define the count of methods in the *dex* file. This variable is 16-bits. $2^{16} = 65536$.

Later Google released a solution to resolve this issue, MultiDex.

MultiDex splits the original *dex* file into multiple *dex* files. The number of methods in each *dex* does not exceed 65536, as shown in Figure 2.32.

FIGURE 2.32 Split *dex* into multiple *dexes*.

classes.dex is also called the main *dex* and is loaded automatically by the app. The app uses *PathClassLoader* to load the main *dex*.

The other *dexes* such as *classes2.dex*, *classes3.dex*, and so on, will be loaded by *DexClassLoader* after the app is launched.

MultiDex is not only used to resolve the 65536 issues but also can be used to improve the performance of the launching app.

The Android system will take a lot of time to load a large *dex* file, so we must take action to reduce the loading time.

We find *classes.dex* doesn't need so many classes. For example, an OTA app has Flight, Hotel, and other modules. We can separate Flight, Hotel, and other modules from *classes.dex* into different *dex* files. We only keep the home page in the *classes.dex*. *classes.dex* will be reduced to a minimal size. Which means we can load *classes.dex* as soon as possible.

We will introduce how to use MultiDex in plug-in solutions in Section 10.4.

2.15 A MUSIC PLAYER APP

Most app developers are not familiar with programming in *Service* and *BroadcastReceiver*. In this section, I will write a music player app to show you how to use these two components.

2.15.1 A Music Player Based on Two *Receivers**

A music player app has many interesting features:

1) Once we open a music player app and play music, if we open another app the music still plays because we use a *Service* to play the music in a background process.

The activity of the music player app is responsible for displaying the information of the current song. There are two buttons "play" and "stop." No matter which button is clicked, *Activity* will send a broadcast to the background *Service* to play or stop the music.

On the other hand, whenever the background *Service* finishes playing music, it will notify the *Activity*, so while the background *Service* plays the next song, it will send another broadcast to the *Activity*. *Activity* will change the name and author of the song.

In this scenario, a music player app requires one *Service* and two *BroadcastReceivers*. The code is shown as follows.

1) Declare Activity and Service in the AndroidManifest.xml:

```
<activity android:name=".MainActivity">
    <intent-filter>
        <action android:name="android.intent.action.
        MAIN" />

        <category android:name="android.intent.
        category.LAUNCHER" />
```

---

* For the example code of this section, please refer to https://github.com/Baobaojianqiang/ReceiverTestBetweenActivityAndService1

```
        </intent-filter>
    </activity>
    <service android:name=".MyService" />
```

2)When the user clicks the play or stop button in *MainActivity*, it will send a broadcast to the *Service*:

```java
public class MainActivity extends Activity {
    TextView tvTitle, tvAuthor;
    ImageButton btnPlay, btnStop;

    Receiver1 receiver1;

    //0x11: stopping; 0x12: playing; 0x13:pausing
    int status = 0x11;

    @Override
    protected void onCreate(Bundle savedInstanceState) {
      super.onCreate(savedInstanceState);
      setContentView(R.layout.activity_main);

      tvTitle = (TextView) findViewById(R.id.tvTitle);
      tvAuthor = (TextView) findViewById(R.id.tvAuthor);

      btnPlay = (ImageButton) this.findViewById
      (R.id.btnPlay);
      btnPlay.setOnClickListener(new View.
      OnClickListener() {
        @Override
        public void onClick(View v) {
          //send message to receiver in Service
          Intent intent = new Intent("UpdateService");
          intent.putExtra("command", 1);
          sendBroadcast(intent);
        }
      });

      btnStop = (ImageButton) this.findViewById
      (R.id.btnStop);
      btnStop.setOnClickListener(new View.
      OnClickListener() {
```

```java
  @Override
  public void onClick(View v) {
    //send message to receiver in Service
    Intent intent = new Intent("UpdateService");
    intent.putExtra("command", 2);
    sendBroadcast(intent);
  }
});

//register receiver in Activity
receiver1 = new Receiver1();
IntentFilter filter = new IntentFilter();
filter.addAction("UpdateActivity");
registerReceiver(receiver1, filter);

//start Service
Intent intent = new Intent(this, MyService.
class);
startService(intent);
}

public class Receiver1 extends BroadcastReceiver {
  @Override
  public void onReceive(Context context, Intent
  intent) {
    status = intent.getIntExtra("status", -1);
    int current = intent.getIntExtra("current",
    -1);
    if (current >= 0) {
      tvTitle.setText(MyMusics.musics[current].
      title);
      tvAuthor.setText(MyMusics.musics[current].
      author);
    }

    switch (status) {
      case 0x11:
        btnPlay.setImageResource(R.drawable.play);
        break;
      case 0x12:
        btnPlay.setImageResource(R.drawable.pause);
        break;
```

```
      case 0x13:
        btnPlay.setImageResource(R.drawable.play);
        break;
      default:
        break;
    }
  }
}
}
```

The relationship between *Receiver1* and *Receiver2* is shown in Figure 2.33.

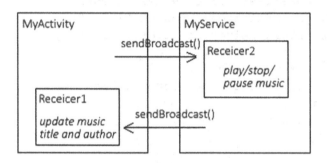

FIGURE 2.33   Two *Receivers* in a music player.

3) In the background, when one song finishes playing and the next song is beginning, *Service* sends a broadcast to *Activity*, shown as follows:

```
public class MyService extends Service {

    Receiver2 receiver2;
    AssetManager am;

    MediaPlayer mPlayer;
    int status = 0x11;
    int current = 0;

    @Override
    public IBinder onBind(Intent intent) {
      return null;
    }

    @Override
    public void onCreate() {
```

```java
am = getAssets();

//register receiver in Service
receiver2 = new Receiver2();
IntentFilter filter = new IntentFilter();
filter.addAction("UpdateService");
registerReceiver(receiver2, filter);

mPlayer = new MediaPlayer();
mPlayer.setOnCompletionListener(new
OnCompletionListener() {
  @Override
  public void onCompletion(MediaPlayer mp) {
    current++;
    if (current >= 3) {
      current = 0;
    }
    prepareAndPlay(MyMusics.musics[current].name);

    //send message to receiver in Activity
    Intent sendIntent = new
    Intent("UpdateActivity");
    sendIntent.putExtra("status", -1);
    sendIntent.putExtra("current", current);
    sendBroadcast(sendIntent);
  }
});
super.onCreate();
}

private void prepareAndPlay(String music) {
  try {
    AssetFileDescriptor afd = am.openFd(music);
    mPlayer.reset();
    mPlayer.setDataSource(afd.getFileDescriptor()
        , afd.getStartOffset()
        , afd.getLength());
    mPlayer.prepare();
    mPlayer.start();
  } catch (IOException e) {
    e.printStackTrace();
  }
}
```

```
public class Receiver2 extends BroadcastReceiver {
  @Override
  public void onReceive(final Context context,
  Intent intent) {
    int command = intent.getIntExtra("command",
    -1);
    switch (command) {
      case 1:
        if (status == 0x11) {
          prepareAndPlay(MyMusics.musics[current].
          name);
          status = 0x12;
        }
        else if (status == 0x12) {
          mPlayer.pause();
          status = 0x13;
        }
        else if (status == 0x13) {
          mPlayer.start();
          status = 0x12;
        }
        break;
      case 2:
        if (status == 0x12 || status == 0x13) {
          mPlayer.stop();
          status = 0x11;
        }
    }

    //send message to receiver in Activity
    Intent sendIntent = new
    Intent("UpdateActivity");
    sendIntent.putExtra("status", status);
    sendIntent.putExtra("current", current);
    sendBroadcast(sendIntent);
  }
}
```

Figure 2.34 shows the relationship between 0x11 (stopping), 0x12 (playing), and 0x13 (pausing).

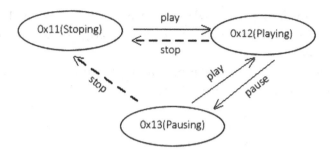

FIGURE 2.34   State machine of a music player.

2.15.2 A Music Player Based on One *Receiver**

In Section 2.15.1, I introduced the first implementation of the music player app. This solution is based on two *Receivers*, one *Receiver* is registered in *Activity*, and the other *Receiver* is registered in *Service*.

In fact, we find almost all the music player apps only have one *Receiver*, which is *Receiver1* introduced in Section 2.15.1 and registered in *Activity*. After a song ends, the background *Service* will send a broadcast to *Receiver1* and update the UI of *Activity*.

On the other hand, we use the method *onBind()* of *Service* to obtain the *Binder* object defined in the *Service*. When the user clicks the button to play or stop the music in *Activity*, it will invoke the methods *play()* or *stop()* of this *Binder* object to operate the background *Service*. Refer to Figure 2.35.

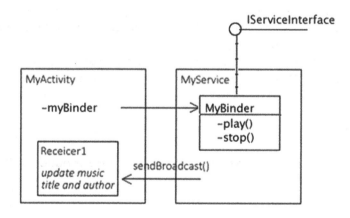

FIGURE 2.35   Class diagram of a music player.

---

* This section example code, please refer to https://github.com/Baobaojianqiang/ReceiverTest BetweenActivityAndService2

In order to decouple the relationship between *Activity* and *Service*, do not use the *MyBinder* object in *Activity* directly, because *MyBinder* is defined in *Service*.

Based on Interface-Oriented programming, we need to create an interface *IServiceInterface*, shown as follows:

```
public interface IServiceInterface {
  public void play();
  public void stop();
}
```

The code in *Activity* is as follows:

```
public class MainActivity extends Activity {
  TextView tvTitle, tvAuthor;
  ImageButton btnPlay, btnStop;

  //0x11: stopping; 0x12: playing; 0x13:pausing
  int status = 0x11;

  Receiver1 receiver1;

  IServiceInterface myService = null;

  ServiceConnection mConnection = new
  ServiceConnection() {
    public void onServiceConnected(ComponentName name,
    IBinder binder) {
      myService = (IServiceInterface) binder;
      Log.e("MainActivity", "onServiceConnected");
    }

    public void onServiceDisconnected(ComponentName
    name) {
      Log.e("MainActivity", "onServiceDisconnected");
    }
  };

  @Override
  protected void onCreate(Bundle savedInstanceState) {
    super.onCreate(savedInstanceState);
```

```
setContentView(R.layout.activity_main);

tvTitle = (TextView) findViewById(R.id.tvTitle);
tvAuthor = (TextView) findViewById(R.id.tvAuthor);

btnPlay = (ImageButton) this.findViewById(R.
id.btnPlay);
btnPlay.setOnClickListener(new View.
OnClickListener() {
  @Override
  public void onClick(View v) {
    myService.play();
  }
});

btnStop = (ImageButton) this.findViewById(R.
id.btnStop);
btnStop.setOnClickListener(new View.
OnClickListener() {
  @Override
  public void onClick(View v) {
    myService.stop();
  }
});

//register receiver in Activity
receiver1 = new Receiver1();
IntentFilter filter = new IntentFilter();
filter.addAction("UpdateActivity");
registerReceiver(receiver1, filter);

//bind Service
Intent intent = new Intent(this, MyService.class);
bindService(intent, mConnection, Context.
BIND_AUTO_CREATE);
}

public class Receiver1 extends BroadcastReceiver {
  @Override
  public void onReceive(Context context, Intent
  intent) {
    status = intent.getIntExtra("status", -1);
```

```
      int current = intent.getIntExtra("current", -1);
      if (current >= 0) {
        tvTitle.setText(MyMusics.musics[current].
        title);
        tvAuthor.setText(MyMusics.musics[current].
        author);
      }

      switch (status) {
        case 0x11:
          btnPlay.setImageResource(R.drawable.play);
          break;
        case 0x12:
          btnPlay.setImageResource(R.drawable.pause);
          break;
        case 0x13:
          btnPlay.setImageResource(R.drawable.play);
          break;
        default:
          break;
      }
    }
  }
}
```

The code in *Service* is as follows:

```
public class MyService extends Service {

  AssetManager am;

  MediaPlayer mPlayer;
  int status = 0x11;
  int current = 0;

  private class MyBinder extends Binder implements
  IServiceInterface {

    @Override
    public void play() {
      if (status == 0x11) {
        prepareAndPlay(MyMusics.musics[current].name);
```

```
      status = 0x12;
    } else if (status == 0x12) {
      mPlayer.pause();
      status = 0x13;
    } else if (status == 0x13) {
      mPlayer.start();
      status = 0x12;
    }

    sendMessageToActivity(status, current);
  }

  @Override
  public void stop() {
    if (status == 0x12 || status == 0x13) {
      mPlayer.stop();
      status = 0x11;
    }

    sendMessageToActivity(status, current);
  }
}

MyBinder myBinder = null;

@Override
public void onCreate() {
  myBinder = new MyBinder();
  am = getAssets();
  mPlayer = new MediaPlayer();
  mPlayer.setOnCompletionListener(new
  OnCompletionListener() {
    @Override
    public void onCompletion(MediaPlayer mp) {
      current++;
      if (current >= 3) {
        current = 0;
      }
      prepareAndPlay(MyMusics.musics[current].name);

      sendMessageToActivity(-1, current);
```

```
      }
    });
    super.onCreate();
  }

  @Override
  public IBinder onBind(Intent intent) {
    return myBinder;
  }

  @Override
  public boolean onUnbind(Intent intent) {
    myBinder = null;
    return super.onUnbind(intent);
  }

  private void sendMessageToActivity(int status1, int
  current1) {
    Intent sendIntent = new Intent("UpdateActivity");
    sendIntent.putExtra("status", status1);
    sendIntent.putExtra("current", current1);
    sendBroadcast(sendIntent);
  }

  private void prepareAndPlay(String music) {
    try {
      AssetFileDescriptor afd = am.openFd(music);
      mPlayer.reset();
      mPlayer.setDataSource(afd.getFileDescriptor()
          , afd.getStartOffset()
          , afd.getLength());
      mPlayer.prepare();
      mPlayer.start();
    } catch (IOException e) {
      e.printStackTrace();
    }
  }
}
```

2.16 SUMMARY

This chapter describes the underlying knowledge of the Android system in detail. I don't list as much code in this chapter, so I have drawn over 30 pictures to illustrate these concepts.

The content of this chapter is the foundation of plug-in techniques. Please study it in detail.

CHAPTER 3

Reflection

THIS CHAPTER INTRODUCES THE most powerful technique in Java: reflection.

The original grammar of reflection is complex and hard to understand, so we can encapsulate it into the class *Utils* for convenience, including the class, constructor, method, and field.

jOOR is a well-known open source library used to encapsulate reflection syntax. However, jOOR is not suitable for the fields marked as *final* in Android. As the author of jOOR noted, jOOR is designed for Java only, and not fully supported in Android.

Another difficulty with reflection is to how to handle a generics class, which is widely used in Android plug-in techniques.

3.1 BASIC REFLECTION*

Reflection consists of the following three techniques:

- Get an object from the class using a string.

- Get all the fields and methods (*public*, *private*, or *static*) of the class.

- Get a generics class.

Compared to the other languages, such as C#, the grammar of reflection in Java is very difficult to understand; we'll introduce the usage of reflection through the three points given above.

---

* For the example code in this section, please refer to *Https://github.com/Baobaojianqiang/ TestReflection*

3.1.1 Get the Class Using a String

3.1.1.1 Get the Class Using a String
We can get class using string, as follows:

```
String str = "abc";
Class c1 = str.getClass();
```

3.1.1.2 Class.forName
Class.forName is widely used in Android.

We can get an object from a string; the string consists of the namespace and its name.

We can also obtain the parent class using *getSuperclass()*:

```
try {
  Class c2 = Class.forName("java.lang.String");
  Class c3 = Class.forName("android.widget.Button");

  //Each Class have this method: getSuperclass().
  Class c5 = c3.getSuperclass(); //Achieve TextView
} catch (ClassNotFoundException e) {
  e.printStackTrace();
}
```

3.1.1.3 Property class
Each class has a property named *class*, as follows:

```
Class c6 = String.class;
Class c7 = java.lang.String.class;
Class c8 = MainActivity.InnerClass.class;
Class c9 = int.class;
Class c10 = int[].class;
```

3.1.1.4 Property TYPE
Primitive types, such as *BOOLEAN*, have a property named *TYPE*, as follows:

```
Class c11 = Boolean.TYPE;
Class c12 = Byte.TYPE;
Class c13 = Character.TYPE;
Class c14 = Short.TYPE;
Class c15 = Integer.TYPE;
```

```
Class c16 = Long.TYPE;
Class c17 = Float.TYPE;
Class c18 = Double.TYPE;
Class c19 = Void.TYPE;
```

In the following chapters, these primitive types are widely used in *Proxy.newProxyinstance()* as parameters.

3.1.2 Get the Property and Method of the Class

3.1.2.1 Get the Constructor of the Class

Now, let's get all the constructors of the class, whether they are marked *private*, *protected*, or *public*, with parameters or not.

For example, *TestClassCtor* has a lot of constructors:

```
public TestClassCtor() {
  name = "baobao";
}

public TestClassCtor(int a) {

}

public TestClassCtor(int a, String b) {
  name = b;
}

private TestClassCtor(int a, double c) {

}
```

3.1.2.1.1 Get All the Constructors of the Class

By using the method *getDeclaredConstructors()*, we get all the constructors of the class, whether *public* or *private*, with parameters or not, and we can iterate each constructor in a loop.

```
TestClass r = new TestClass();
Class temp = r.getClass();
String className = temp.getName(); // Gets the name of
                                      the specified class

Log.v("baobao", "Gets all constructors of the class,
no matter public or private----------------------");
```

```
//Gets all constructors of the class, no matter public
or private
try {
    Constructor[] theConstructors = temp.
    getDeclaredConstructors();

    for (int i = 0; i < theConstructors.length; i++) {
      int mod = theConstructors[i].getModifiers();
      // Get labels and it's name of a constructor
      Log.v("baobao", Modifier.toString(mod) + " " +
      className + "(");

        //Gets the collection parameters of the
        specified constructor
      Class[] parameterTypes = theConstructors[i].
      getParameterTypes();
      for (int j = 0; j < parameterTypes.length; j++) {
      // Printout parameter list
        Log.v("baobao", parameterTypes[j].getName());
        if (parameterTypes.length > j + 1) {
          Log.v("baobao", ", ");
        }
      }
      Log.v("baobao", ")");
    }
  } catch (Exception e) {
    e.printStackTrace();
  }
```

If we want to get all the *public* constructors of the class, but without *private* constructors, we can use *GetConstructors()* instead of *getDeclaredConstructors()*.

3.1.2.1.2 Get a Constructor of a Class
Get a constructor without parameters:

```
Constructor c1 = temp.getDeclaredConstructor();
```

Get a constructor with an integer parameter:

```
Class[] p2 = {int.class};
Constructor c2 = temp.getDeclaredConstructor(p2);
```

Get a constructor with two parameters—one is an integer and the other is a string:

```
Class[] p3 = {int.class, String.class};
Constructor c3 = temp.getDeclaredConstructor(p3);
```

It is important to get the constructor of a class as it is a crucial step in the following process:

- Get a class using a string.
- Get the constructor of this class.
- Get an instance of the class by executing the constructor.
- Get all the fields and methods of the instance using reflection.

3.1.2.1.3 Call a Constructor
We can use the method *newInstance()* to get an instance of the class:

```
Class r = Class.forName("jianqiang.com.testreflection.
TestClassCtor");

//Containing parameters
Class[] p3 = {int.class, String.class};
Constructor ctor = r.getDeclaredConstructor(p3);
Object obj = ctor.newInstance(1, "bjq");

// without parameters
Constructor ctor2 = r.getDeclaredConstructor();
Object obj2 = ctor2.newInstance();
```

3.1.2.2 Invoke a Private Method of the Class
TestClassCtor has a private method named *doSomething*:

```
private String doSomething(String d) {
  Log.v("baobao", "TestClassCtor, doSomething");
  return "abcd";
}
```

Invoke this method by reflection, as follows:

```
Class r = Class.forName("jianqiang.com.
testreflection.TestClassCtor");
Class[] p3 = {int.class, String.class};
Constructor ctor = r.getDeclaredConstructor(p3);
Object obj = ctor.newInstance(1, "bjq");

// Call a private method like this:
Class[] p4 = {String.class};
Method = r.getDeclaredMethod("doSomething", p4); //
Gets the specified method in the specified class
method.setAccessible(true);

Object argList[] = {"jianqiang"}; //As a parameter
needed in the method
Object result = method.invoke(obj, argList);
```

3.1.2.3 Invoke a Private and Static Method of the Class
TestClassCtor has a private and static method named *work*:

```
private static void work() {
  Log.v("baobao", "TestClassCtor, work");
}
```

Invoke this method by reflection, as follows:

```
Class r = Class.forName("jianqiang.com.testreflection.
TestClassCtor");
// To Call this method, do as following :
Method = r.getDeclaredMethod("work"); //Get the
specified method in the specified class
method.setAccessible(true);
method.invoke(null);
```

3.1.2.4 Get a Private Field of the Class and Modify Its Value
TestClassCtor has a private field named "*name*":

```
public class TestClassCtor {
  private String name;

  public String getName() {
    return name;
  }
}
```

Get this field and modify its value, as follows:

```
//Get a class instance from its name by inflection
Class r = Class.forName("jianqiang.com.testreflection.
TestClassCtor");
Class[] p3 = {int.class, String.class};
Constructor ctor = r.getDeclaredConstructor(p3);
Object obj = ctor.newInstance(1, "bjq");

//Get the private field: name
Field = r.getDeclaredField("name");
field.setAccessible(true);

Object fieldObject = field.get(obj);

//Effective only to obj
field.set(obj, "jianqiang1982");
```

The modification in the above code is only effective for the current object, and if we create another instance of *TestClassCtor*, the value of its "name" field will be null, rather than "jianqiang1982."

```
TestClassCtor = new TestClassCtor(100);
testClassCtor.getName(); // Return Null instead of
jianqiang1982
```

3.1.2.5 Get the Private Static Field of the Class and Modify Its Value
TestClassCtor has a static and private field named "*address*":

```
public class TestClassCtor {
  private static String address;
}
```

Get this static private field and modify its value, as follows:

```
//Get a class instance from its name by inflection
Class r = Class.forName("jianqiang.com.testreflection.
TestClassCtor");

//Get the private static field: address
Field = r.getDeclaredField("address");
field.setAccessible(true);
```

```
//the parameter is null when the field is static
Object fieldObject = field.get(null);

field.set(fieldObject, "ABCD");

// As static field, modification is effective once
modified to another instance of the class
TestClassCtor.printAddress();
```

Once we use reflection to modify the value of a static field, it will take effect forever. Next time we visit this static field, it will be the new value.

3.1.3 Generics and *Singleton<T>*

Generics, such as the class *Singleton*, are widely used in the source code of Android.

```
public abstract class Singleton<T> {
  private T mInstance;

  protected abstract T create();

  public final T get() {
    synchronized (this) {
      if (mInstance == null) {
        mInstance = create();
      }
      return mInstance;
    }
  }
}
```

Singleton is a generics class, and we can get the *mInstance* field of *Singleton*, as follows:

```
Class<?> Singleton = Class.forName("jianqiang.com.
testreflection.Singleton");
Field mInstanceField = singleton.
getDeclaredField("mInstance");
mInstanceField.setAccessible(true);
```

Singleton is also an abstract class, which has an abstract method named *create*.

Let's look at the *ActivityManagerNative* class; it is also called *AMN* for short.

AMN is usually associated with *Singleton* as follows:

```
public class AMN {
  private static final Singleton<ClassB2Interface>
  gDefault = new Singleton<ClassB2Interface>() {
    protected ClassB2Interface create() {
      ClassB2 b2 = new ClassB2();
      b2.id = 2;
      return b2;
    }
  };

  static public ClassB2Interface getDefault() {
    return gDefault.get();
  }
}
```

The method *getDefault* is a static private field of *AMN*, and its return type is *Singleton*, so it must implement the *create* method, and return an instance of *ClassB2*.

Now, let's get this object by reflection.

First, we get the field *gDefault* of *AMN*, which is a static and private field:

```
Class<?> activityManagerNativeClass = Class.
forName("jianqiang.com.testreflection.AMN");
Field gDefaultField = activityManagerNativeClass.
getDeclaredField("gDefault");
gDefaultField.setAccessible(true);
Object gDefault = gDefaultField.get(null);
```

Second, we get the object *rawB2Object* from *gDefault*, as follows:

```
// The original B2 object inside the gDefault object
of AMN
Object rawB2Object = mInstanceField.get(gDefault);
```

rawB2Object is an instance of *ClassB2*.

However, we find that *rawB2object* is not the object that we need, so we convert it into the object *proxy*, whose type is *ClassB2Mock*.

ClassB2Mock is a dynamic-proxy for *rawB2Object*. We use the method *Proxy.newProxyInstance()* to create this relationship between *ClassB2Mock* and *rawB2Object*, as follows:

```
// Create a proxy object for this instance
Classb2mock, and then replace this field, then our
agents can help work
Class<?> classB2Interface = Class.forName("jianqiang.
com.testreflection.ClassB2Interface");
Object proxy = Proxy.newProxyInstance(
  Thread.currentThread().getContextClassLoader(),
  new Class<?>[] { classB2Interface },
  new ClassB2Mock(rawB2Object));
mInstanceField.set(gDefault, proxy);
```

The last line of code above is to set the field *mInstance* of *gDefault* to *proxy*, which is created by the method *Proxy.newProxyInstance()*.

We call the above process a hook. After a hook has taken place, *AMN.getDefault().doSomething()* will execute the logic of the *ClassB2Mock*.

The behavior of the *ActivityManagerService* in the source code of Android is the same as we discuss in this chapter. We use *ClassB2* and *ClassB2Mock* to simulate a real scenario in the Android system.

3.2 jOOR*

The grammar in the above example is based on Java syntax, which is very complex and inconvenient.

It is much more convenient to replace it with a simple and object-oriented grammar, so we have jOOR.[†]

jOOR only has two classes, *Reflect.java* and *ReflectException.java*, so we can directly drag them to our project instead of relying on *Gradle*.

Reflect.java is the soul of jOOR. It has six core methods:

- *on*: wrap a class or an object. When we wrap a class, the parameter may be a class type or a string, as follows:

```
Reflect r1 = on(Object.class);
```

* Code sample: https//github.com/Baobaojianqiang/TestReflection2
† jOOR address: https://github.com/jOOQ/jOOR

- *create*: invoke the constructor of the class or object wrapped using "*on*" syntax, with parameters or not, as follows:

```
Reflect = on("android.widget.Button").create();
```

- *call*: call a method, take the method's name and parameters as its parameters. If the calling method has a return value, we could use *get()* to get its value, as follows:

```
reflect.call("doSomething", "param1").get()
```

- *get*: get a field or the return value of the method, support type conversion. It is usually used with the *call* method, as follows:

```
Object obj1 = obj.get("name");
```

- *set*: set the value of a field of the object, as follows:

```
obj.set("name", "jianqiang");
```

We use jOOR to refactor the sample code seen in Section 3.1.

3.2.1 Get a Class from a String

3.2.1.1 Get a Class from a String

We can get a class using a string, as follows:

```
String str = "abc";
Class c1 = str.getClass();
```

3.2.1.2 Get a Class by Using on and get

When we are using jOOR, we generally import its *Reflect.on* method so that we can use "*on*" directly in the code to make the code simpler.

```
import static jianqiang.com.testreflection.joor.
Reflect.on;

// The following three lines of code are equivalent
Reflect r1 = on(Object.class);
Reflect r2 = on("java.lang.Object");
Reflect r3 = on("java.lang.Object", ClassLoader.
getSystemClassLoader());
```

```
// The following two lines of code are equivalent,
achieving an Object and getting Object.class
Object o1 = on(Object.class).<Object>get();
Object o2 = on("java.lang.Object").get();

String j2 = on((Object)"abc").get();
int j3 = on(1).get();

// Equivalent to Class.forName()
try {
    Class j4 = on("android.widget.Button").type();
} catch (ReflectException e) {
    e.printStackTrace();
}
```

3.2.2 Get the Property and Method of a Class

3.2.2.1 Get a Constructor of a Class

Let's get all the constructors of the class, whether marked *private*, *protected*, or *public*, with parameters or not.

jOOR doesn't support getting or invoking a constructor directly, but we can use the method *create()* to instantiate an object:

```
TestClassCtor r = new TestClassCtor();
Class temp = r.getClass();
String className = temp.getName(); // Get the name of
                                      the specified class

//Public Constructor
Object obj = on(temp).create().get();      //without
                                           parameters
Object obj2 = on(temp).create(1, "abc").get();
//Having parameters

//Private Constructor
TestClassCtor obj3 = on(TestClassCtor.class).create
(1, 1.1).get();
String a = obj3.getName();
```

3.2.2.2 Get the Private Method of the Class

Get the private instance method of the class and invoke this method:

```
// The following four lines of code are used to get an
object
```

```
TestClassCtor r = new TestClassCtor();
Class temp = r.getClass();
Reflect = on(temp).create();

//Invoke a method
String a1 = reflect.call("doSomething", "param1").
get();
```

3.2.2.3 Get the Private and Static Method of the Class

Get the private and static method of the class and invoke this method:

```
// The following four lines of code are used to get an
object
TestClassCtor r = new TestClassCtor();
Class temp = r.getClass();
Reflect reflect = on(temp).create();

//Invoke a static method
on(TestClassCtor.class).call("work").get();
```

3.2.2.4 Get the Private Field of the Class

Get the private field of the class and modify its value:

```
Reflect obj = on("jianqiang.com.testreflection.
TestClassCtor").create(1, 1.1);
obj.set("name", "jianqiang");
Object obj1 = obj.get("name");
```

3.2.2.5 Get the Private and Static Field of the Class

Get the private and static field of the class and modify its value:

```
on("jianqiang.com.testreflection.TestClassCtor").
set("address", "avcccc");
Object obj2 = on("jianqiang.com.testreflection.
TestClassCtor").get("address");
```

3.2.3 Generics and Singleton<T>

It's easy to handle generics in jOOR:

```
// Obtaining gDefault which is static and Singleton in
AMN
Object gDefault = on("jianqiang.com.testreflection.
AMN").get("gDefault");
```

```
// gDefault is an android.util.Singleton object; we
obtain the mInstance field in Singleton.
// mInstance is the original ClassB2Interface object
Object mInstance = on(gDefault).get("mInstance");

// Create a proxy object for this instance
Classb2mock, and then replace this field, then our
agents can help work
Class<?> classB2Interface = on("jianqiang.com.
testreflection.ClassB2Interface").type();
Object proxy = Proxy.newProxyInstance(
  Thread.currentThread().getContextClassLoader(),
  new Class<?>[] { classB2Interface },
  new ClassB2Mock(mInstance));

on(gDefault).set("mInstance", proxy);
```

jOOR doesn't fully support Android, for example, for reflecting a *final* field.

There are two *final* fields in the *User* class: one is marked *final* and *static* and the other is marked *final* only.

```
public class User {
  private final static int userId = 3;
  private final String name = "baobao";
}
```

Let's use jOOR to reflect these two fields:

```
//Final field
Reflect obj = on("jianqiang.com.testreflection.User").
create();
obj.set("name", "jianqiang");
Object newObj = obj.get("name");

//Static field
Reflect obj2 = on("jianqiang.com.testreflection.
User");
obj2.set("userId", "123");
Object newObj2 = obj2.get("userId");
```

It will throw up a *NoSuchFieldException* when we execute the *set* method in the code above. The root cause of this is that the *set* method in jOOR will try to reflect the *modifier* field of the *Field* class. The *modifier* field exists in the Java environment but doesn't exist in the Android environment.

3.3 ENCAPSULATED CLASSES OF THE BASIC REFLECTION*

Considering the limitations of jOOR, we need to find another way to support *final*. We try to encapsulate the basic Java reflection grammar into a lot of simple methods by ourselves, as follows:

- Reflect a constructor and invoke it.

- Call a static method.

- Call an instance method.

- Get a field and set its value.

- Handle generics.

3.3.1 Get a Constructor

Define the *createObject* method in the *RefInvoke* class:

```
public static Object createObject(String className,
Class[] pareTyples, Object[] pareVaules) {
    try {
        Class r = Class.forName(className);
        Constructor ctor = r.getDeclaredConstructor
        (pareTyples);
        ctor.setAccessible(true);
        return ctor.newInstance(pareVaules);
    } catch (Exception e) {
        e.printStackTrace();
    }

    return null;
}
```

---

* Code sample: https//github.com/Baobaojianqiang/TestReflection3

Use the *createObject* method as follows:

```
Class r = Class.forName(className);

//With parameters
Class[] p3 = {int.class, String.class};
Object[] v3 = {1, "bjq"};
Object obj = RefInvoke.createObject(className, p3, v3);

// without parameters
Object obj2 = RefInvoke.createObject(className, null,
null);
```

3.3.2 Invoke Instance Methods

Define the *invokeInstanceMethod* method in the *RefInvoke* class:

```
public static Object invokeInstanceMethod(Object obj,
String methodName, Class[] pareTyples, Object[]
pareVaules) {
   if(obj == null)
     return null;

   try {
     //Call a private method
     //Get the specified method in the specified class
     Method = obj.getClass().
     getDeclaredMethod(methodName, pareTyples);
     method.setAccessible(true);
     return method.invoke(obj, pareVaules);

   } catch (Exception e) {
     e.printStackTrace();
   }

   return null;
}
```

Use the *invokeInstanceMethod* method as follows:

```
Class[] p3 = {};
Object[] v3 = {};
RefInvoke.invokeStaticMethod(className, "work", p3, v3);
```

3.3.3 Invoke Static Methods

Define the *invokeStaticMethod* method in the *RefInvoke* class:

```
public static Object invokeStaticMethod(String
className, String method_name, Class[] pareTyples,
Object[] pareVaules) {
    try {
      Class obj_class = Class.forName(className);
      Method = obj_class.getDeclaredMethod(method_name,
      pareTyples);
      method.setAccessible(true);
      return method.invoke(null, pareVaules);
    } catch (Exception e) {
      e.printStackTrace();
    }

    return null;
}
```

Use the *invokeStaticMethod* method as follows:

```
Class[] p4 = {String.class};
Object[] v4 = {"jianqiang"};
Object result = RefInvoke.invokeInstanceMethod(obj,
"doSomething", p4, v4);
```

3.3.4 Get the Field of the Class and Set Its Value

Define the *getFieldObject* and *setFieldObject* methods in the *RefInvoke* class:

```
public static Object getFieldObject(String className,
Object obj, String fieldName) {
    try {
      Class obj_class = Class.forName(className);
      Field = obj_class.getDeclaredField(fieldName);
      field.setAccessible(true);
      return field.get(obj);
    } catch (Exception e) {
      e.printStackTrace();
    }

    return null;
}
```

```
public static void setFieldObject(String classname,
Object obj, String fieldName, Object fieldVaule) {
    try {
        Class obj_class = Class.forName(classname);
        Field field = obj_class.
        getDeclaredField(fieldName);
        field.setAccessible(true);
        field.set(obj, fieldVaule);
    } catch (Exception e) {
        e.printStackTrace();
    }
}
```

Use the *getFieldObject* and *setFieldObject* methods as follows:

```
//Get a field
Object fieldObject = RefInvoke.
getFieldObject(className, obj, "name");
RefInvoke.setFieldObject(className, obj, "name",
"jianqiang1982");

//Get a field labeled static
Object fieldObject = RefInvoke.
getFieldObject(className, null, "address");
RefInvoke.setFieldObject(className, null, "address",
"ABCD");
```

3.3.5 Handle Generics

Up until now, we have encapsulated five methods in the *RefInvoke* class; now it's time to introduce how to implement generics, as follows:

```
//Obtaining gDefault which is static and Singleton
in AMN
Object gDefault = RefInvoke.getFieldObject("jianqiang.
com.testreflection.AMN", null, "gDefault");

// gDefault is an android.util.Singleton object;
we obtain the mInstance field in Singleton.
Object rawB2Object = RefInvoke.getFieldObject(
    "jianqiang.com.testreflection.Singleton",
    gDefault, "mInstance");
```

```
// Create a proxy object for this instance
ClassB2Mock, and then replace this field, then our
proxy can help Class<?> classB2Interface = Class.
forName("jianqiang.com.testreflection.
ClassB2Interface");
Object proxy = Proxy.newProxyInstance(
    Thread.currentThread().getContextClassLoader(),
    new Class<?>[] { classB2Interface },
    new ClassB2Mock(rawB2Object));

// Replaced mInstance in Singleton with Proxy
RefInvoke.setFieldObject("jianqiang.com.
testreflection.Singleton", gDefault, "mInstance",
proxy);
```

This code is simple to understand and use in our project.

3.4 FURTHER ENCAPSULATION OF THE REFLECTION*

In Section 3.3, we introduced a new class named *RefInvoke*, which has five methods. Then we encapsulated a complex reflection logic into these five methods, but in practice we find it is not convenient in some scenarios. Let's resolve these small issues in this section.

3.4.1 Reflect a Method with Only One Parameter or without Parameters

Sometimes we create an object using a constructor without parameters, but we still need to set its parameter to *null* in the *createObject* method, as follows:

```
Class r = Class.forName(className);

//With parameters
Class[] p3 = {int.class, String.class};
Object[] v3 = {1, "bjq"};
Object obj = RefInvoke.createObject(className, p3, v3);

//without parameters
Object obj2 = RefInvoke.createObject(className, null,
null);
```

---

* Code sample: https://github.com/Baobaojianqiang/TestReflection4

Sometimes we find the constructor has only one parameter, but we still need convert this parameter into an array, or it doesn't make sense, as follows:

```
//With only one parameter
Class[] p3 = {int.class};
Object[] v3 = {1};
Object obj = RefInvoke.createObject(className, p3, v3);
```

We want to make the code simple, so we supply a series of overload methods *createObject()* , as follows:

```
// without parameters
public static Object createObject(String className) {
    Class[] pareTyples = new Class[]{};
    Object[] pareVaules = new Object[]{};

    try {
        Class r = Class.forName(className);
        return createObject(r, pareTyples, pareVaules);
    } catch (ClassNotFoundException e) {
        e.printStackTrace();
    }

    return null;
}

//with one parameter
public static Object createObject(String className,
Class pareTyple, Object pareVaule) {
    Class[] pareTyples = new Class[]{ pareTyple };
    Object[] pareVaules = new Object[]{ pareVaule };

    try {
        Class r = Class.forName(className);
        return createObject(r, pareTyples, pareVaules);
    } catch (ClassNotFoundException e) {
        e.printStackTrace();
    }

    return null;
}
```

```
//multiple parameters
public static Object createObject(String className,
Class[] pareTyples, Object[] pareVaules) {
    try {
        Class r = Class.forName(className);
        return createObject(r, pareTyples, pareVaules);
    } catch (ClassNotFoundException e) {
        e.printStackTrace();
    }

    return null;
}

//multiple parameters
public static Object createObject(Class clazz, Class[]
pareTyples, Object[] pareVaules) {
    try {
        Constructor ctor = clazz.getDeclaredConstructor
        (pareTyples);
        ctor.setAccessible(true);
        return ctor.newInstance(pareVaules);
    } catch (Exception e) {
        e.printStackTrace();
    }

    return null;
}
```

Now we can create an object with less code, as follows:

```
//With only one parameter
Object obj = RefInvoke.createObject(className, int.
class, 1);
//without parameters
Object obj2 = RefInvoke.createObject(className);
```

Actually, the constructor is a method, which means we also need supply a series of overload methods *invokeStaticMethod()* and *invokeInstanceMethod()*.

3.4.2 Replace String with Class Type

Up until now, we have obtained a class by using its full name, as follows:

```
public static Object createObject(String className,
Class[] pareTyples, Object[] pareVaules) {
   try {
      Class r = Class.forName(className);
      Constructor ctor = r.getConstructor(pareTyples);
      return ctor.newInstance(pareVaules);
   } catch (Exception e) {
      e.printStackTrace();
   }

   return null;
}
```

However, sometimes we have the class type rather than a string, so we don't need to use *Class.forName(className)* anymore; we can use the class type directly, as follows:

```
//multiple parameters
public static Object createObject(String className,
Class[] pareTyples, Object[] pareVaules) {
   try {
     Class r = Class.forName(className);
     return createObject(r, pareTyples, pareVaules);
   } catch (ClassNotFoundException e) {
     e.printStackTrace();
   }

   return null;
}

//multiple parameters
public static Object createObject(Class clazz, Class[]
pareTyples, Object[] pareVaules) {
   try {
     Constructor ctor = clazz.getConstructor(pareTyples);
     return ctor.newInstance(pareVaules);
   } catch (Exception e) {
     e.printStackTrace();
   }

   return null;
}
```

All the methods in *RefInvoke* have two forms, one is a string, and the other one is a class type.*

3.4.3 Differences between the Static and Instance Fields

When we get or set a field by reflection, we find that the grammar is nearly the same, whether it's a static field or instance field. There is only one difference between these two fields, as follows:

```
//Get a common Field
Object fieldObject = RefInvoke.
getFieldObject(className, obj, "name");
RefInvoke.setFieldObject(className, obj, "name",
"jianqiang1982");

//Get a static field
Object fieldObject = RefInvoke.
getFieldObject(className, null, "address");
RefInvoke.setFieldObject(className, null, "address",
"ABCD");
```

The static field doesn't need the parameter *obj*, so we set it to *null* instead.

We don't want to see the *null* value in our code, so we encapsulate the reflection logic of the static method into two new methods, *getstaticfieldobject* and *setstaticfieldobject*, as follows:

```
public static Object getStaticFieldObject(String
className, String fieldName) {
    return getFieldObject(className, null, fieldName);
}

public static void setStaticFieldObject(String
classname, String fieldName, Object fieldVaule) {
    setFieldObject(classname, null, fieldName,
    fieldVaule);
}
```

Now we can write a simple code as follows:

```
Object fieldObject = RefInvoke.
getFieldObject(className, null, "address");
RefInvoke.setStaticFieldObject(className, "address",
"ABCD");
```

---

* Sample code: https://github.com/BaoBaoJianqiang/TestReflection4

3.4.4 Optimization of the Field Reflection

Let's continue to demonstrate the encapsulation of the instance field, as follows:

```
public static Object getFieldObject(Class clazz,
Object obj, String fieldName) {
    try {
      Field = clazz.getDeclaredField(fieldName);
      field.setAccessible(true);
      return field.get(obj);
    } catch (Exception e) {
      e.printStackTrace();
    }

    return null;
}
```

We find that the type of the second parameter *obj* is mostly equal to the first parameter *clazz*, which means we can omit the first parameter *clazz* from the method *getFieldObject*, as follows:

```
public static Object getFieldObject(Object obj, String
fieldName) {
    return getFieldObject(obj.getClass(), obj,
    fieldName);
}

public static void setFieldObject(Object obj, String
fieldName, Object fieldVaule) {
    setFieldObject(obj.getClass(), obj, fieldName,
    fieldVaule);
}
```

Now we can write a simple code as follows:

```
Object fieldObject = RefInvoke.
getFieldObject(className, "address");
RefInvoke.setStaticFieldObject(className, "address",
"ABCD");
```

Sometimes the type of *obj* is not equal to *clazz*, and we are unable to omit the first parameter *clazz* anymore.

3.5 SUMMARY

This chapter shows three ways to use reflection:

1) Use traditional reflection grammar.

2) Use jOOR.

3) Encapsulate the basic reflection grammar.

jOOR doesn't fully support Android, and we usually use the third solution in our Android projects.

In this book, we use the third solution in all the demos.

Proxy Pattern

To understand what a plug-in is, we must first look at the proxy pattern, which is one of 23 software design patterns. There are two ways to implement a proxy pattern, one is *Static-Proxy*, and the other is *Dynamic-Proxy*. In plug-in technology, these two patterns are applied to hook *Instrumentation* and *AMN*. I will introduce them in detail in this chapter.

4.1 WHAT IS A PROXY PATTERN?

Proxy patterns appear everywhere in our software projects. Every class with the suffix "*Proxy*" usually makes use of the proxy pattern, such as *ActivityManagerProxy*

The definition of a proxy pattern in GOF23 is: a wrapper or agent object that is called by the client to access the real serving object behind the scenes.

A UML diagram of the proxy pattern is shown in Figure 4.1.

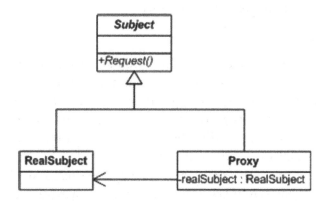

FIGURE 4.1   Class diagram of a proxy pattern.

In Figure 4.1, *realSubject* and *Proxy* are both subclasses of *Subject*. A *Proxy* class has a field named *realSubject* whose type is *realSubject*. The *Proxy* class also has a method *Request*; its implementation is to invoke *realSubject*'s *Request* method. The code is shown as follows:

```
abstract public class Subject {
  abstract public void Request();
}

public class realSubject : Subject {
  public override void Request()
  {
    //Do Something
  }
}

public class Proxy : Subject {
  private realSubject realSubject;

  public override void Request()
  {
    //Below is the key statement
    realSubject.Request();
  }
}
```

In the next section, I will describe the usage of the proxy pattern, and you will discover for yourselves its power.

4.1.1 Remote Proxy

WebService, a very popular technology around 2004, is implemented in *Remote Proxy*. We could transfer data from the Java server to the .NET server. *WebService* creates mapping between *XML* and the entity class in Java or C#. First, the entity written in Java is converted to *XML*, and then the *XML* data is transferred to the .NET server, and then *XML* is converted to the entity in C# in the .NET server.

Now, *WebService* techniques have already been replaced with JSON. Because *XML* in *WebService* is too heavy, we usually have to create a lot of files like DTD or Schema to define the format of the *XML*. But JSON is lighter than *XML*.

Let's go back to the Android system. *AIDL* is implemented in *Remote Proxy*, as shown in Figure 4.2.

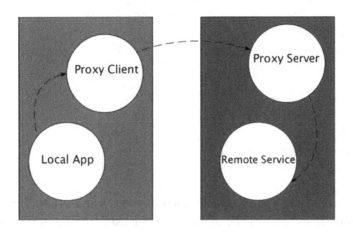

FIGURE 4.2 *AIDL* with a proxy pattern.

As we talked about in previous chapters, *AIDL* has a *Client* and a *Server*. For example, we can define the method *sum* in *AIDL*,

```
sum(int a, int b)
```

AIDL will help us when generating Java source code automatically; the implementation of the *add* method in its *Proxy* class is below:

```
private static class Proxy implements com.lypeer.
ipcclient.Caculator {
  @Override
  public int add(int a, int b) throws android.
  os.RemoteException {
     android.os.Parcel _data = android.os.Parcel.
     obtain();
     android.os.Parcel _reply = android.os.Parcel.
     obtain();
     int _result;
     try {
        _data.writeInterfaceToken(DESCRIPTOR);
        _data.writeInt(a);
        _data.writeInt(b);
```

```
        mRemote.transact(Stub.TRANSACTION_add, _data,
        _reply, 0);
        _reply.readException();
        _result = _reply.readInt();
    } finally {
        _reply.recycle();
        _data.recycle();
    }
    return _result;
}

//ignore some code...
}
```

The method *add* will write variables *a* and *b* to *_data*, and then send *_data* and *_reply* to another endpoint of *AIDL* using the method *transact* of the object *mRemote*. The object *_reply* is a callback function; it will bring back the result. This is a typical proxy pattern, as shown in Figure 4.3.

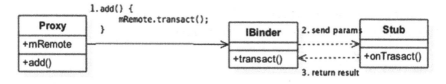

FIGURE 4.3   A proxy pattern with *Proxy* and *IBinder*.

The other endpoint of *AIDL* will read the values from *_data* by calling on the method *onTransact* in the class *Stub*, then calculate the result, and return the proxy via the object *_reply*.

4.1.2 Write Log

The class *Class1* has a method *doSomething*; we must write a log before the method *doSomething* is executed. Commonly, we add a statement to write a log at the beginning of the method *doSomething*.

After we have learned the proxy pattern, we design the class *Class1Proxy* as shown in Figure 4.4, with the following code:

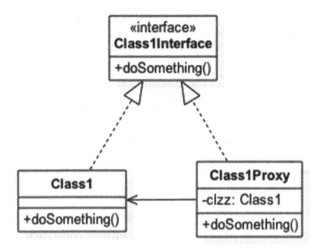

FIGURE 4.4   Implementation of a *Static-Proxy.*

```
public class Class1Proxy implements Class1Interface {
  Class1 clzz = new Class1();

  @Override
  public void doSomething() {
    System.out.println("Begin log");
    clzz.doSomething();
    System.out.println("End log");
  }
}
```

Then we use the class *Class1Proxy* instead of *Class1,*

```
Class1Proxy proxy = new Class1Proxy();
proxy.doSomething();
```

4.2 *STATIC-PROXY* AND *DYNAMIC-PROXY**

Continue with the sample above, since it was a proxy pattern, but there're some problems with it. We need to write a *Proxy* class for each origin class, and the number of classes will be increased quickly. But the logic is nearly the same. We call this *Static-Proxy.*

Dynamic-Proxy can resolve this problem. Now let's have a look at an important piece of syntax: *Proxy.newProxyInstance().*

---

* Sample code: https://github.com/BaoBaoJianqiang/InvocationHandler

```
static Object newProxyInstance(
  ClassLoader loader,
  Class<?>[] interfaces,
  InvocationHandler h)
```

There are three parameters in the above function:

- *Loader*: the *ClassLoader* of target class *Class1*.

- *Interfaces*: the interface type of the target class *Class1*; here it is *Class1Interface*.

- The third parameter is an implementation of *InvocationHandler*; we will inject the target object of *Class1* with its constructor.

```
Class1Interface class1 = new Class1();
Class1Interface class1Proxy = (Class1Interface) Proxy.
newProxyInstance(
       class1.getClass().getClassLoader(),
       class1.getClass().getInterfaces(),
       new InvocationHandlerForTest(class1));
class1Proxy.doSomething();
```

Through the method *Proxy.newProxyInstance()*, we could create a new instance of *Class1Interface*, which is a *class1Proxy*.

class1Proxy.doSomething will call the method *invoke()* of *InvocationHandlerForTest*.

The second parameter of the method *invoke*, named *method*, is actually the method *doSomething* of *Class1*.

We can write a log before or after *method.invoke()*, as follows:

```
public class InvocationHandlerForTest implements
InvocationHandler {
  private Object target;

  public InvocationHandlerForTest(Object target) {
    this.target = target;
  }

  @Override
  public Object invoke(Object o, Method method,
  Object[] objects) throws Throwable {
```

```
    System.out.println("Begin Log");
    Object obj = method.invoke(target, objects);
    System.out.println("End Log");
    return obj;
  }
}
```

Focused on *method.invoke(target, objects)*, *target* is *Class1* and *objects* is a parameter of the method *doSomething* of *Class1*.

Then if we invoke the method *doSomething* of *class1Proxy*, the method *doSomething* of *Class1* will be *invoked*.

The method *Proxy.newProxyInstance()* could be applied to any instance of the interface, and adds a new function to the original object, so we call it a *Dynamic-Proxy*.

In the plug-in technique, an object created by *Proxy.newProxyInstance()* would be used to replace the original one, we call this technique a hook.

4.3 A HOOK ON THE *AMN**

This section is an extension of Chapter 3, *Reflection*.

In Chapter 3, we defined two classes *AMN* and *Singleton*. In fact, these two classes represent the implementation of *ActivityManagerNative* of the Android source code, but the implementation in Chapter 3 was simple.

The communication between Android components (*Activity, Service,* and so on) and the *AMS* are based on *AMN* or *AMP*, as we mentioned in Chapter 2.

For example, an app invokes the method *startActivity*:

```
ActivityManagerNative.getDefault().startActivity()
```

And sends a message by the method *sendBroadcast()*:

```
ActivityManagerNative.getDefault().broadcastIntent()
```

In Chapter 3, we introduced reflection. Now we understand that *ActivityManagerNative.getDefault()* invokes the method *get* of the class *Singleton*, and returns the field *mInstance* of *Singleton*.

We can replace *mInstance* with our own logic, for example, to print a log. The code is shown as follows:

---

* Code sample: https://github.com/BaoBaoJianqiang/hookAMS

```
public final class HookHelper {

  public static void hookActivityManager() {
    try {
      // To get the Singleton of AMN, gDefault, it's a
      static property
      Object gDefault = RefInvoke.
      getStaticFieldObject("android.app.
      ActivityManagerNative", "gDefault");

      // gDefault is an android.util.Singleton object;
      we get mInstance property of it, it's type of
      IActivityManager
      Object rawIActivityManager = RefInvoke.
      getFieldObject(
          "android.util.Singleton",
          gDefault, "mInstance");


      //Create the proxy object of
      iActivityManagerInterface, then replace the
      property, and let it do something
      Class<?> iActivityManagerInterface = Class.
      forName("android.app.IActivityManager");
      Object proxy = Proxy.newProxyInstance(
          Thread.currentThread().
          getContextClassLoader(),
          new Class<?>[] { iActivityManagerInterface },
          new HookHandler(rawIActivityManager));

      //Replace mInstance of Singleton with proxy
      RefInvoke.setFieldObject("android.util.
      Singleton", gDefault, "mInstance", proxy);

    } catch (Exception e) {
      throw new RuntimeException("Hook Failed", e);
    }
  }
}
```

The logic in *HookHandler* is simple, it will print logs before the original method is executed.

```
class HookHandler implements InvocationHandler {

  private static final String TAG = "HookHandler";

  private Object mBase;

  public HookHandler(Object base) {
    mBase = base;
  }

  @Override
  public Object invoke(Object proxy, Method method,
  Object[] args) throws Throwable {
    Log.d(TAG, "hey, baby; you are hooked!!");
    Log.d(TAG, "method:" + method.getName() + " called
    with args:" + Arrays.toString(args));

    return method.invoke(mBase, args);
  }
}
```

4.4 A HOOK ON THE *PMS**

The *PMS* is one of the services in the Android system; we can't hook it directly. In fact, we can only hook its proxy object within the Android system process; it's an object of the class *PackageManager*. We can also find it in a lot of classes.

ActivityThread has a field *sPackageManager*.

ApplicationPackageManager has a field *mPM*.

We try to hook these two fields and print some logs.

```
public static void hookPackageManager(Context context) {
  try {
// Get the global ActivityThread object
    Object currentActivityThread = RefInvoke.
    getStaticFieldObject("android.app.
    ActivityThread", "currentActivityThread");

    // Get origin sPackageManager from ActivityThread
    Object sPackageManager = RefInvoke.getFieldObject
    (currentActivityThread, "sPackageManager");
```

```
// Prepare the proxy object to replace the
original object
Class<?> iPackageManagerInterface = Class.
forName("android.content.pm.IPackageManager");
Object proxy = Proxy.newProxyInstance(iPackageMan
agerInterface.getClassLoader(),
    new Class<?>[] { iPackageManagerInterface },
    new HookHandler(sPackageManager));

// 1. Replace sPackageManager from ActivityThread
RefInvoke.setFieldObject(sPackageManager,
"sPackageManager", proxy);

// 2. Replace mPm from ApplicationPackageManager
PackageManager pm = context.getPackageManager();
RefInvoke.setFieldObject(pm, "mPM", proxy);

    } catch (Exception e) {
        throw new RuntimeException("hook failed", e);
    }
}
```

4.5 SUMMARY

This chapter is examined two kinds of proxy patterns in the Android system, and looked at a powerful method called *Proxy.newProxyInstance()*.

In this chapter, we tried to hook *AMN* and the *PMS* to print some logs. In Chapter 5, we will hook the method *startActivity* to launch an *Activity* which has not been declared in the *AndroidManifest.xml*.

Hooking *startActivity()*

I N CHAPTER 4, WE talked about how to hook the *Proxy* of the *AMS* and the *PMS* during app processes.

In this chapter, with experience of hooking the method *startActivity()*, we can start an *Activity* without it being declared in the *AndroidManifest. xml*.

5.1 INVOKE *STARTACTIVITY()* IN TWO WAYS

One way to invoke the method *startActivity()* in *Activity*:

```
Intent intent = new Intent(MainActivity.this,
SecondActivity.class);
startActivity(intent);
```

Another way is to invoke the method *startActivity()* of the *Context*; we execute *getApplicationContext()* to get an instance of *Context*, shown as follows:

```
Intent intent = new Intent(MainActivity.this,
SecondActivity.class);
getApplicationContext().startActivity(intent);
```

But the essence of these two ways is the same, as shown in Figure 5.1 and 5.2. Compare these two figures: we find the steps before *AMN/AMP* are different; the other steps are the same.

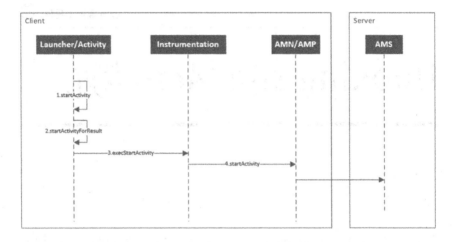

FIGURE 5.1 Sequence diagram of *startActivity()* in *Activity*.

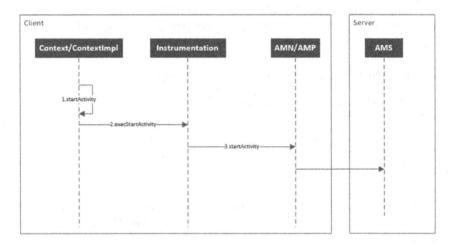

FIGURE 5.2 Sequence diagram of *startActivity()* in *Context*.

5.2 HOOKING *STARTACTIVITY()* OF THE *ACTIVITY*

When we invoke the method *startActivity()* of *Activity*, we want to write our own logic, for example, printing some logs.

Press the button in *Activity1* to navigate to *Activity2*; the whole process is very long. I introduced this process in Section 2.6. Now let's focus on the first step and the last step of this process.

- first, *Activity1* notifies the *AMS* to navigate to *Activity2*

- second, the *AMS* notifies the app to navigate to *Activity2*

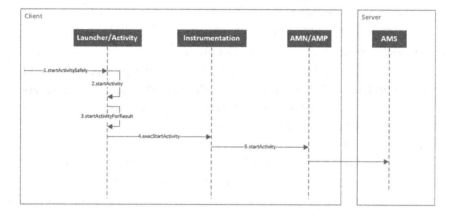

FIGURE 5.3   Process from the app to the *AMS*.

According to Figure 5.3, we can hook in the following three places:

- the method *startActivityForResult()* of *Activity*
- the field *mInstrumentation* of *Activity*
- the method *getDefault()* of *AMN*

According to Figure 5.4, we can hook in the following two places:

- the field *mCallback* of *H*
- the field *mInstrumentation* of *ActivityThread*, to hook its methods *newActivity()* and *callActivityOnCreate()*

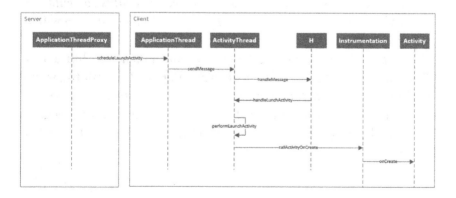

FIGURE 5.4   Process from the *AMS* to the app.

5.2.1 Solution 1: Hooking the Method
startActivityForResult of *Activity*

We usually create a class named *BaseActivity* in the app, and all the *Activities* defined need to inherit from it, then we can override the method *startActivityForResult()* in *BaseActivity*. Because the method *startActivity()* invokes the method *startActivityForResult()* in *Activity*, we can override the method *startActivityForResult()*.

5.2.2 Solution 2: Hooking the Field *mInstrumentation* of *Activity**

According to Figure 5.3, there is a field *mInstrumentation* inside *Activity*, and the method *startActivityForResult()* of *Activity* invokes the method *execStartActivity()* of the field *mInstrumentation*, shown as follows:

```
public class Activity extends ContextThemeWrapper {
  private Instrumentation mInstrumentation;

  public void startActivityForResult(Intent intent,
  int requestCode, @Nullable Bundle options) {
    //Omit some code
    Instrumentation.ActivityResult ar =
      mInstrumentation.execStartActivity(
        this, mMainThread.getApplicationThread(),
        mToken, this,
        intent, requestCode, options);
  }
}
```

The field *mInstrumentation* of *Activity* is *private*; we need to obtain this field by reflection and replace it with an object of *EvilInstrumentation*. This means the method *execStartActivity()* of *EvilInstrumentation* will be invoked, which is referred to Figure 5.5.

EvilInstrumentation inherits from *Instrumentation* and has a field *mBase*; *mBase* is also a type of *Instrumentation*, and we call the wrapper of *Instrumentation* by the name *EvilInstrumentation*. When we call the method *execStartActivity()* of *EvilInstrumentation*, we write logs in this method, and then invoke the method *execStartActivity()* of *mBase* (Figure 5.6).

---

* Code sample: https://github.com/BaoBaoJianqiang/hook11

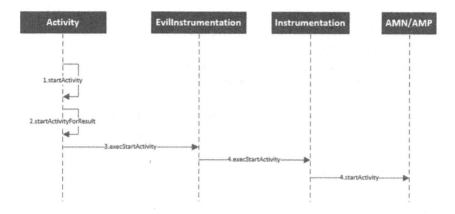

FIGURE 5.5   Invoke the method *execStartActivity()* of *EvilInstrumentation*.

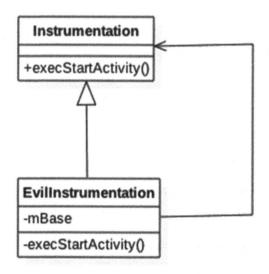

FIGURE 5.6   *EvilInstrumentation* extends *Instrumentation*.

1) The logic in *MainActivity*; hooking *Instrumentation* with the method *onCreate()*:

```
public class MainActivity extends Activity {

@Override
protected void onCreate(Bundle savedInstanceState) {
  super.onCreate(savedInstanceState);
```

```
Instrumentation mInstrumentation = (Instrumentation)
RefInvoke.getFieldOjbect(Activity.class, this,
"mInstrumentation");
Instrumentation evilInstrumentation = new EvilInst
rumentation(mInstrumentation);

RefInvoke.setFieldOjbect(Activity.class, this,
"mInstrumentation", evilInstrumentation);

Button tv = new Button(this);
tv.setText("Test Page");
setContentView(tv);

tv.setOnClickListener(new View.OnClickListener() {
  @Override
  public void onClick(View v) {
    Intent intent = new Intent(MainActivity.this,
    SecondActivity.class);
    startActivity(intent);
  }
});
  }
}
```

2) The logic in *EvilInstrumentation*, which overrides the method *execStartActivity*():

```
public class EvilInstrumentation extends
Instrumentation {

  private static final String TAG =
  "EvilInstrumentation";

  // saved the original object in ActivityThread
  Instrumentation mBase;

  public EvilInstrumentation(Instrumentation base) {
    mBase = base;
  }

  public ActivityResult execStartActivity(
```

```
    Context who, IBinder contextThread, IBinder
    token, Activity target,
    Intent intent, int requestCode, Bundle options) {

  Log.d(TAG, "XXX visited!");

  // Start calling the original method, call it or
  leave it, but if you don't, all the
  startActivity() will expire.
  // Since this method is hidden, you need to use
  reflection to invoke.
  Class[] p1 = {Context.class, IBinder.class,
      IBinder.class, Activity.class,
      Intent.class, int.class, Bundle.class};
  Object[] v1 = {who, contextThread, token, target,
      intent, requestCode, options};
  return (ActivityResult) RefInvoke.
  invokeInstanceMethod(
      mBase, "execStartActivity", p1, v1);
  }
}
```

This mechanism only takes effect in the current *Activity*, and it replaces the current value of the field *mInstrumentation* with an *EvilInstrumentation* object. When we navigate to another *Activity*, it doesn't work.

We can implement this logic in *BaseActivity*. Because we require that all the *Activities* inherit from *BaseActivity*, each activity has the field *mInstrumentation* which is replaced with an *EvilInstrumentation* object.

5.2.3 Solution 3: Hooking the Method *getDefault()* of *AMN**

In Figure 5.3, the method *execStartActivity*() of *Instrumentation* invokes the method *startActivity()* of *AMN*, shown as follows:

```
public class Instrumentation {

  public ActivityResult execStartActivity(
     Context who, IBinder contextThread, IBinder
     token, Activity target,
     Intent intent, int requestCode, Bundle options) {
```

---

* Code sample: https://github.com/BaoBaoJianqiang/hook12

```
//Omit some code
int result = ActivityManagerNative.getDefault()
    .startActivity(whoThread, who.
    getBasePackageName(), intent,
        intent.resolveTypeIfNeeded(who.
        getContentResolver()),
        token, target != null ? target.mEmbeddedID :
        null,
        requestCode, 0, null, options);
    }
}
```

The method *getDefault()* of *AMN*, described in Section 4.3, returns an *IActivityManager* object.

IActivityManager is an interface, we can use the method *Proxy.new-ProxyInstance()* to replace the *IActivityManager* object with a *MockClass1* object, shown as follows:

```
public class AMSHookHelper {
  public static final String EXTRA_TARGET_INTENT =
  "extra_target_intent";

  public static void hookAMN() throws
  ClassNotFoundException,
      NoSuchMethodException, InvocationTargetException,
      IllegalAccessException, NoSuchFieldException {

    //Gets the gDefault singleton of AMN, which is
    final static
    Object gDefault = RefInvoke.
    getStaticFieldOjbect("android.app.
    ActivityManagerNative", "gDefault");

    // gDefault is an android.util.Singleton<T>
    object; We fetch the mInstance inside the
    singleton
    Object mInstance = RefInvoke.
    getFieldOjbect("android.util.Singleton", gDefault,
    "mInstance");

    // Create a proxy object MockClass1 for this
    object, then replace the field and let our proxy
    object do the work
```

```
Class<?> classB2Interface = Class.
forName("android.app.IActivityManager");
Object proxy = Proxy.newProxyInstance(
    Thread.currentThread().getContextClassLoader(),
    new Class<?>[] { classB2Interface },
    new MockClass1(mInstance));

//Change the mInstance field of gDefault to proxy
RefInvoke.setFieldOjbect("android.util.Singleton",
gDefault, "mInstance", proxy);
    }
}
```

ClassMockClass1 need follow two points:

1) Implement the interface *InvocationHandler*.

2) Intercept the method *startActivity()* and print a log before executing it.

The code is shown as follows:

```
class MockClass1 implements InvocationHandler {

  private static final String TAG = "MockClass1";

  Object mBase;

  public MockClass1(Object base) {
    mBase = base;
  }

  @Override
  public Object invoke(Object proxy, Method method,
  Object[] args) throws Throwable {

    if ("startActivity".equals(method.getName())) {

      Log.e("bao", method.getName());

      return method.invoke(mBase, args);
    }
```

```
      return method.invoke(mBase, args);
   }
}
```

Then we can invoke the method *hookAMN()* of *AMSHookHelper* in the lifecycle method *attachBaseContext()* of *MainActivity*. Click the button in *MainActivity* and enter into *MockClass1*.

attachBaseContext() is a lifecycle method of *MainActivity*, and it's executed earlier than the lifecycle method *onCreate()*. We need to hook it as soon as possible to make sure the hook takes effect in the following *Activities*.

There is a scenario in which the user clicks a button in the web browser and navigates to a detailed page of the app. *MainActivity* has no chance to show, which means the hook logic of *attachBaseContext()* won't be executed.

In all plug-in frameworks, we move the hooking logic into the method *attachBaseContext()* of *Application*. As we introduced in Chapter 2, *Application* is created when the app process is created, so the method *attachBaseContext()* of *Application* is executed before each *Activity* is born.

In this chapter, we choose to hook in the *attachBaseContext()* of *MainActivity* for simplicity.

```
public class MainActivity extends Activity {

  @Override
  protected void attachBaseContext(Context newBase) {
    super.attachBaseContext(newBase);

    try {
      AMSHookHelper.hookAMN();
    } catch (Throwable throwable) {
      throw new RuntimeException("hook failed",
      throwable);
    }
  }

  @Override
  protected void onCreate(Bundle savedInstanceState) {
    super.onCreate(savedInstanceState);
    Button = new Button(this);
```

```
button.setText("start TargetActivity");

button.setOnClickListener(new View.
OnClickListener() {
  @Override
  public void onClick(View v) {
    Intent intent = new Intent(MainActivity.this,
    TargetActivity.class);
    startActivity(intent);
  }
});
setContentView(button);
  }
}
```

5.2.4 Solution 4: Hooking the Field *mCallback* of *H**

The whole process of *startActivity()* is long. We have introduced three ways to hook in the preceding sections; each of them occurs in the process which sends messages from the app to AMS.

From now on, let's talk about the process which sends a message from the *AMS* to the app.

In Figure 5.7, *ActivityThread* has a field *mH*. The type of *mH* is *H*. Class *H* inherits from *Handler*, meaning class *H* inherits the function of dispatching messages.

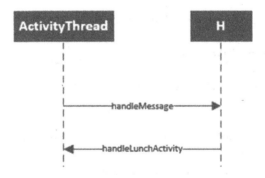

FIGURE 5.7   Communication between *ActivityThread* and *H*.

Handler has an object named *mCallback*. The type of *mCallback* is *Callback*. Because *H* inherits from *Handler*, the object *mH* has a field *mCallback* (Figure 5.8).

---

* Code sample: https://github.com/BaoBaoJianqiang/hook13

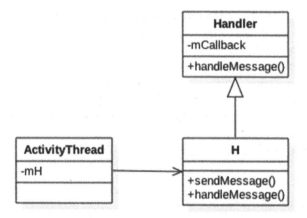

FIGURE 5.8   Relationship between *ActivityThread* and *H*.

The *AMS* sends a message to the app to launch an *Activity*, and then *ActivityThread* invokes the method *sendMessage()* of *mH*. The value in the message to launch an *Activity* is 100 (*LAUNCH_ACTIVITY*).

The method *sendMessage()* of *mH* invokes the method *handleMessage()* of *mCallback* to send the message.

As a result, we can hook the field *mCallback*, to replace it with a *Callback* object. App developers can use the class *Callback* directly.

Now let's hook the field *mCallback* of *H* to intercept its method *handleMessage()*.

1) The implementation of *HookHelper*:

```
public class HookHelper {

  public static void attachBaseContext() throws
  Exception {

    // First, obtain the object for current
    ActivityThread
    Object currentActivityThread = RefInvoke.
    getStaticFieldOjbect("android.app.ActivityThread",
    "sCurrentActivityThread");

    // Because there is ActivityThread in a process,
    so we obtain the field mH
    Handler mH = (Handler) RefInvoke.
    getFieldOjbect("android.app.ActivityThread",
    currentActivityThread, "mH");
```

```
  //replace the mCallback to new MockClass2(mH)
  RefInvoke.setFieldOjbect(Handler.class, mH,
  "mCallback", new MockClass2(mH));
  }
}
```

2) The implementation of *MockClass2*:

```
public class MockClass2 implements Handler.Callback {

  Handler mBase;

  public MockClass2(Handler base) {
    mBase = base;
  }

  @Override
  public boolean handleMessage(Message msg) {

    switch (msg.what) {
      // "LAUNCH_ACTIVITY" inside ActivityThread value
      100
      // The best way is by using reflection, which is
      to use hard coding for simplicity
      case 100:
        handleLaunchActivity(msg);
        break;
    }

    mBase.handleMessage(msg);
    return true;
  }

  private void handleLaunchActivity(Message msg) {
    // get the TargetActivity;

    Object obj = msg.obj;

    Log.d("baobao", obj.toString());
  }
}
```

Now, we might question why we don't hook the field *mH* of *Activity Thread* directly?

Let's have a quick look at the proceeding hooking demos. We find they are separated into two different solutions, one is *Static-Proxy*, and the other one is *Dynamic-Proxy*:

- *Static-Proxy*: there are two classes suitable for this solution, one is *Handler.Callback*, the other one is *Instrumentation*. Only these two classes are exposed to app developers.

- *Dynamic-Proxy*: there are two interfaces suitable for this solution, *IPackageManager*, and *IActivityManager*. The method *Proxy.newProxyInstance()* only supports the interface.

Look at the field *mH* of *ActivityThread*. The type of *mH* is *H*. *H* is an internal class, and it's not open to app developers, so app developers cannot create an object which inherits from *H* to replace the field *mH*. Also, *H* is not an interface; we can't use *Proxy.newProxyInstance()* to replace it.

Unfortunately, most of the classes in the Android system are not open to app developers.

5.2.5 Solution 5: Hooking *Instrumentation* Again*

From Figure 5.4, let's focus on the last two steps; I simplify it as shown in Figure 5.9.

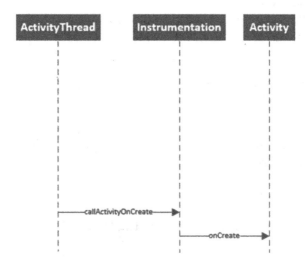

FIGURE 5.9 Launch a new *Activity*.

---

* Code sample: https://github.com/BaoBaoJianqiang/hook14

According to Section 5.2.2, there is a field *mInstrumentation* in *Activity*, and we replace it with our own *EvilInstrumentation* object.

ActivityThread has a field *mInstrumentation* too. *ActivityThread* invokes the method *newActivity()* of *mInstrumentation* to generate an *Activity* object and then invoke the method *callActivityOnCreate()* of *mInstrumentation* to start this new *Activity*.

By the way, we can also hook the field *mInstrumentation* of *ActivityThread* and replace it with our own *EvilInstrumentation*, but this time we're intercepting the method *newActivity()* and *callActivityOnCreate()* of *Instrumentation*.

1) The logic in *HookHelper* to replace the field *mInstrumentation* of *ActivityThread*:

```
public class HookHelper {

  public static void attachContext() throws Exception{
    // First, obtain the object of current
    ActivityThread
    Object currentActivityThread = RefInvoke.
    invokeStaticMethod("android.app.ActivityThread",
    "currentActivityThread");

    // obtain original the mInstrumentation field
    Instrumentation mInstrumentation =
    (Instrumentation) RefInvoke.
    getFieldOjbect("android.app.ActivityThread"
    ,currentActivityThread, "mInstrumentation");

    // create a proxy object
    Instrumentation evilInstrumentation = new EvilInst
    rumentation(mInstrumentation);

    // Replace
    RefInvoke.setFieldOjbect("android.app.
    ActivityThread" ,currentActivityThread,
    "mInstrumentation", evilInstrumentation);
  }
}
```

2) The logic in *EvilInstrumentation* to print logs in the method *newActivity()* and *callActivityOnCreate()*:

```
public class EvilInstrumentation extends
Instrumentation {

  private static final String TAG =
  "EvilInstrumentation";

  // save the original object inside ActivityThread
  Instrumentation mBase;

  public EvilInstrumentation(Instrumentation base) {
    mBase = base;
  }

  public Activity newActivity(ClassLoader cl, String
  className,
              Intent intent)
      throws InstantiationException,
      IllegalAccessException,
      ClassNotFoundException {

    Log.d(TAG, "XXX visited!");

    return mBase.newActivity(cl, className, intent);
  }

  public void callActivityOnCreate(Activity activity,
  Bundle bundle) {

    Log.d(TAG, "XXX visited!");

    Class[] p1 = {Activity.class, Bundle.class};
    Object[] v1 = {activity, bundle};
    RefInvoke.invokeInstanceMethod(
        mBase, "callActivityOnCreate", p1, v1);
  }
}
```

5.3 HOOKING THE METHOD STARTACTIVITY OF CONTEXT

There are two ways to launch a new *Activity*. We have introduced a way which is widely used in app development: to invoke the method *startActivity()* of the *Activity*.

This section introduces another way to launch an *Activity*, the method *startActivity()* of *Context*.

5.3.1 Solution 6: Hooking the Field *mInstrumentation* of *ActivityThread**

We can use the method *startActivity()* of *Context* to launch a new *Activity*. We get a *Context* object through the method *getApplicationContext()*.

```
Intent intent = new Intent(MainActivity.this,
SecondActivity.class);
getApplicationContext().startActivity(intent);
```

In the method *startActivity()* of *Context*, it invokes the method *startActivity()* of *ContextImpl*.

The method *startActivity()* of *ContextImpl* invokes the method *execStartActivity()* of the field *mInstrumentation* of *ActivityThread*, shown as follows:

```
class ContextImpl extends Context {
  @Override
  public void startActivity(Intent intent, Bundle
  options) {
    //Omit some code
    mMainThread.getInstrumentation().
    execStartActivity(
        getOuterContext(), mMainThread.
        getApplicationThread(), null,
        (Activity) null, intent, -1, options);
  }
}
```

As a result, we can hook the field *mInstrumentation* of *ActivityThread* and intercept the method *execStartActivity()* of *Instrumentation*. The code is shown as follows:

---

* Code sample: https://github.com/BaoBaoJianqiang/hook15

1) The logic of *HookHelper*, which we introduced in Section 5.2.5:

```
public class HookHelper {

  public static void attachContext() throws Exception{
    // First, obtain the object of current
    ActivityThread
    Object currentActivityThread = RefInvoke.
    invokeStaticMethod("android.app.ActivityThread",
    "currentActivityThread");

    // obtain original mInstrumentation field
    Instrumentation mInstrumentation =
    (Instrumentation) RefInvoke.
    getFieldOjbect("android.app.ActivityThread",
    currentActivityThread, "mInstrumentation");

    // create a proxy object
    Instrumentation evilInstrumentation = new EvilInst
    rumentation(mInstrumentation);

    // transformation
    RefInvoke.setFieldOjbect("android.app.
    ActivityThread", currentActivityThread,
    "mInstrumentation", evilInstrumentation);
  }
}
```

2) The logic of *EvilInstrumentation*, which we introduced in Section 5.2.2:

```
public class EvilInstrumentation extends
Instrumentation {

  private static final String TAG =
  "EvilInstrumentation";

  // save the original object inside ActivityThread
  Instrumentation mBase;

  public EvilInstrumentation(Instrumentation base) {
    mBase = base;
  }
```

```
public ActivityResult execStartActivity(
    Context who, IBinder contextThread, IBinder
    token, Activity target,
    Intent, int requestCode, Bundle options) {

  Log.d(TAG, "XXX visited!");

  Class[] p1 = {Context.class, IBinder.class,
      IBinder.class, Activity.class,
      Intent.class, int.class, Bundle.class};
  Object[] v1 = {who, contextThread, token, target,
      intent, requestCode, options};
  return (ActivityResult) RefInvoke.
  invokeInstanceMethod(
      mBase, "execStartActivity", p1, v1);
  }
}
```

5.3.2 Which Solution Is the Best?

The process of hooking the method *startActivity()* of *Context*, is to get an *Instrumentation* object from *ActivityThread*, and then execute the method *execStartActivity()* of the *Instrumentation*, and finally invoke the method *AMN.getDefault().startActivity()*.

Up until now, we have introduced six solutions, but we find that solution 3, described in Section 5.2.3 is the best one, which hooks the method *getDefault()* of *AMN*.

Therefore, solution 1, 2, and 6 are not widely used in plug-in programming. All the plug-in frameworks use solution 3.

5.4 LAUNCH AN ACTIVITY NOT DECLARED IN ANDROIDMANIFEST.XML

In the traditional Android app development, an app can't launch an *Activity* not declared in the *AndroidManifest.xml*.

Now let's look at this limitation in Android.

5.4.1 How to Hook *AMN*

Let's have a look at the process of the navigation from one *Activity* to another *Activity*, as shown in Figure 5.10.

When an app launches a new *Activity* that is not declared in the *AndroidManifest.xml*, the *AMS* will perform a check on whether

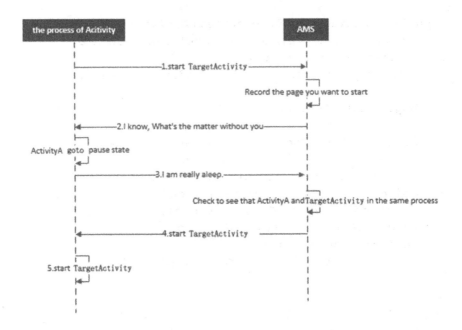

FIGURE 5.10 Navigation from one *Activity* to another *Activity*.

the new *Activity* is declared in the *AndroidManifest.xml* in step 2 of Figure 5.10. If the *Activity* is not declared, the *AMS* will throw an *ActivityNotFoundException.*

We need to create a false impression for the *AMS* that the new *Activity* exists in *AndroidManifest.xml*.

But we can't modify the hooking logic in the *AMS* process; the *AMS* is a system process, and we have no permissions to modify it; otherwise, every app would modify the logic in the *AMS*, and Android system would not be secure. For example, we write an app to change the original behavior of Clipboard. In this app, whatever the content we cut or copy, the content we paste is the same string. But it takes effect only in this app. When we open another app, the Clipboard recovers its original behavior, and the paste content is the content we cut or copied before.

We can't hook the *AMS* in step 2; we can only write a hook login in the app in step 1 (before checking) and 5 (launch a new *Activity*).

The hooking solution is that we declare an empty *Activity* which has no logic. It's only a placeholder, so we name it *StubActivity.*

In step 1 of Figure 5.10, we replace the *TargetActivity* with *StubActivity* declared in the *AndroidManifest.xml* before we send the *TargetActivity* information to the *AMS*. The *AMS* will check the validation of *StubActivity* rather *TargetActivity*. In the process of replacement, we store the information of the original *TargetActivity* in a *Bundle*.

In step 5 of Figure 5.10, when the *AMS* notifies the app to launch *StubActivity*, we replace *StubActivity* with the original *TargetActivity*. The information of the *TargetActivity* is stored in a *Bundle* (Figure 5.11).

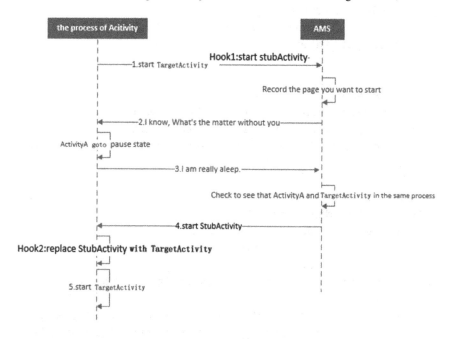

FIGURE 5.11   Hook the launching process of *Activity*.

5.4.2 First Half of the Hook*

Up until now, we have mainly used hooking with the method *startActivity()*. Before the app sends the information of the *Activity* to the *AMS*, we have three points to hook, as follows:

- Hook the field *mInstrumentation* of *Activity*, which is applicable to the method *startActivity()* of *Activity*.

- Hook the field *mInstrumentation* of *ActivityThread*, which is applicable to the method *startActivity()* of *Context*.

- Hooking on *AMN* is applied to the methods *startActivity()* of *Activity* and *Context*.

---

* Code sample: https://github.com/BaoBaoJianqiang/hook31

We have introduced the first two points in the previous sections; now, we're going to hook *AMN* and replace *TargetActivity* with *StubActivity*.

1) The code of *AMSHookHelper*:

```
public class AMSHookHelper {
  public static final String EXTRA_TARGET_INTENT =
"extra_target_intent";

  /**
   * Hook AMS
   * The main thing you do is "temporarily replace the
   Activity you really want to start with a
   StubActivity declared in androidmanifest.xml," and
   hook AMS
   */
  public static void hookAMN() throws
  ClassNotFoundException,
     NoSuchMethodException, InvocationTargetException,
     IllegalAccessException, NoSuchFieldException {

    //Obtained the singleton gDefault of AMN, which is
    static final
    Object gDefault = RefInvoke.getStaticFieldOjbect
    ("android.app.ActivityManagerNative", "gDefault");

    // The getDefault is an object of android.util.
    Singleton<T>,we take out the mIntance field inside
    the singleton
    Object mInstance = RefInvoke.getFieldOjbect
    ("android.util.Singleton", gDefault, "mInstance");

    // Create a proxy object MockClass1 for this
    object, then replace the field and let our proxy
    object do the work
    Class<?> classB2Interface = Class.
    forName("android.app.IActivityManager");
    Object proxy = Proxy.newProxyInstance(
       Thread.currentThread().getContextClassLoader(),
       new Class<?>[] { classB2Interface },
       new MockClass1(mInstance));
```

```
  //The field mIntance of the gDefault changes to
  proxy
  RefInvoke.setFieldOjbect("android.util.Singleton",
  gDefault, "mInstance", proxy);
  }
}
```

2) The code of *MockClass1*:

```
class MockClass1 implements InvocationHandler {

  private static final String TAG = "MockClass1";

  Object mBase;

  public MockClass1(Object base) {
    mBase = base;
  }

  @Override
  public Object invoke(Object proxy, Method method,
  Object[] args) throws Throwable {

    Log.e("bao", method.getName());

    if ("startActivity".equals(method.getName())) {
      // Intercept the method only
      // Replace the parameters, whatever you want; You
      can even replace the original Activity and start
      another Activity

      // Find the first Intent object in the parameter
      Intent raw;
      int index = 0;

      for (int i = 0; i < args.length; i++) {
        if (args[i] instanceof Intent) {
          index = i;
          break;
        }
      }
```

```
    raw = (Intent) args[index];

    Intent newIntent = new Intent();

    // The package name of the stunt Activity, which
    is our own package name
    String stubPackage = raw.getComponent().
    getPackageName();

    // So we're going to temporarily replace the
    starting Activity with the StubActivity
    ComponentName componentName = new
    ComponentName(stubPackage, StubActivity.class.
    getName());
    newIntent.setComponent(componentName);

    // Save the TargetActivity that we originally
    started
    newIntent.putExtra(AMSHookHelper.EXTRA_TARGET_
    INTENT, raw);

    // Replace Intent to cheat AMS
    args[index] = newIntent;

    Log.d(TAG, "hook success");
    return method.invoke(mBase, args);

  }

  return method.invoke(mBase, args);
 }
}
```

We use *MockClass1* to intercept the method *startActivity()*, and do the following work step by step:

1) Get the original *Intent* from the parameters of the method *startActivity()*;

2) Create a *NewIntent* object to launch *StubActivity*;

3) Save the original *Intent* in the *NewIntent* object;

4) Replace the original *Intent* with the *NewIntent* object.

The original *Intent* stored in the *NewIntent* object will be used in the next section.

5.4.3 Second Half of the Hook: Hooking the Field *mCallback* of *H**

After the app cheats the *AMS* into launching *StubActivity*, the *AMS* will notify the app to start *StubActivity* in step 4 of Figure 5.10. We have no permissions to modify the *AMS* process. We can only modify the app process in step 5 of Figure 5.10, for example, changing *StubActivity* to *TargetActivity*.

In Section 5.2.4 and 5.2.5, we have introduced two kinds of hooking techniques:

- Hooking the field *mCallback* of *H*;

- Hooking the field *mInstrumentation* of *ActivityThread*.

The solution in this section is based on hooking the field *mCallback* of *H*, shown as follows:

1) *AMSHookHelper* is the same as solution 4 introduced in Section 5.2.4.

```
public class HookHelper {
  public static final String EXTRA_TARGET_INTENT =
  "extra_target_intent";

  public static void attachBaseContext() throws
  Exception {

    // First, obtain the object of current
    ActivityThread
    Object currentActivityThread = RefInvoke.
    getStaticFieldOjbect("android.app.ActivityThread",
    "sCurrentActivityThread");

    // we obtain the field mH
```

---

* Code sample: https://github.com/BaoBaoJianqiang/hook31

```
   Handler mH = (Handler) RefInvoke.
   getFieldOjbect("android.app.ActivityThread",
   currentActivityThread, "mH");

   //replace the mCallback to new MockClass2(mH)
   RefInvoke.setFieldOjbect(Handler.class, mH,
   "mCallback", new MockClass2(mH));
   }
}
```

2) *MockClass2* intercepts messages with the value 100 (which stands for the method *startActivity()*) and replaces *StubActivity* with *TargetActivity*:

```
class MockClass2 implements Handler.Callback {

  Handler mBase;

  public MockClass2(Handler base) {
    mBase = base;
  }

  @Override
  public boolean handleMessage(Message msg) {

    switch (msg.what) {
      // "LAUNCH_ACTIVITY" inside ActivityThread value
      100
      // The best way is by using reflection, which is
      to use hard coding for simplicity
      case 100:
        handleLaunchActivity(msg);
        break;
    }

    mBase.handleMessage(msg);
    return true;
  }

  private void handleLaunchActivity(Message msg) {
    // In this case, take out the TargetActivity in
    simple way;;
    Object obj = msg.obj;
```

```
Intent intent = (Intent) RefInvoke.
getFieldOjbect(obj.getClass(), obj, "intent");

Intent targetIntent = intent.getParcelableExtra(AM
SHookHelper.EXTRA_TARGET_INTENT);
intent.setComponent(targetIntent.getComponent());
  }
}
```

5.4.4 Second Half of the Hook: Hooking the *mInstrumentation* Field of *ActivityThread**

1) *HookHelper* hooks the field *mInstrumentation* of *ActivityThread*, and has the same implementation as introduced in Section 5.2.5:

```
public class HookHelper {
  public static final String EXTRA_TARGET_INTENT =
  "extra_target_intent";

  public static void attachContext() throws Exception{
    // First, obtain the object of current
    ActivityThread
    Object currentActivityThread = RefInvoke.
    invokeStaticMethod("android.app.ActivityThread",
    "currentActivityThread");

    // obtain original the mInstrumentation field
    Instrumentation mInstrumentation =
    (Instrumentation) RefInvoke.
    getFieldOjbect("android.app.ActivityThread",
    currentActivityThread, "mInstrumentation");

    // create a proxy object
    Instrumentation evilInstrumentation = new EvilInst
    rumentation(mInstrumentation);

    // transformation
    RefInvoke.setFieldOjbect("android.app.
    ActivityThread" ,currentActivityThread,
    "mInstrumentation", evilInstrumentation);
  }
}
```

---

* Code sample: https://github.com/BaoBaoJianqiang/hook32

2) As we discussed in Section 5.2.5, we can hook the field *mInstrumentation* of *ActivityThread*, and intercept two methods in *Instrumentation*, shown as follows:

- Activity newActivity(ClassLoader cl, String className, Intent intent)

- void callActivityOnCreate(Activity activity, Bundle icicle, PersistableBundle persistentState)

We can't get an *Intent* object from the method *callActivityOnCreate()*, but we can get an *Intent* object in the parameters of the method *newActivity()*, so we choose to intercept *newActivity()*:

```
public class EvilInstrumentation extends
Instrumentation {

  private static final String TAG =
  "EvilInstrumentation";

  // The package name of the Stub Activity, which is
  our own package name
  String packageName = "jianqiang.com.hook1";

  // saved the original object in ActivityThread
  Instrumentation mBase;

  public EvilInstrumentation(Instrumentation base) {
    mBase = base;
  }

  public Activity newActivity(ClassLoader cl, String
  className,
              Intent intent)
    throws InstantiationException,
    IllegalAccessException,
    ClassNotFoundException {

  Intent rawIntent = intent.getParcelableExtra
  (HookHelper.EXTRA_TARGET_INTENT);
  if(rawIntent == null) {
```

```
    return mBase.newActivity(cl, className, intent);
  }

  String newClassName = rawIntent.getComponent().
  getClassName();
  return mBase.newActivity(cl, newClassName,
  rawIntent);
  }
}
```

Note that we need to judge whether the object *rawIntent* is null or not in the method *newActivity()*. If the object *rawIntent* is null, it means the process is normal, and we don't hook.

5.5 SUMMARY

This chapter introduced the process of the method *startActivity()*, and looked for the places suitable to hook with this method.

As a result, there are six hooking solutions.

Based on these six hooking solutions, we can launch an *Activity* not declared in the *AndroidManifest.xml*.

Try to understand all six solutions in this chapter as they are widely used in plug-in apps.

CHAPTER 6

The Basic Concepts of Plug-In Techniques

IN THIS CHAPTER, WE begin to introduce plug-in techniques. This chapter will talk about how to load classes into a plug-in app.

We find it's not convenient to load classes in plug-ins by reflection each time, the code is ugly, that's why we use Interface-Oriented programming in plug-in frameworks.

Last, we introduce how to use Android Studio to debug plug-in apps.

6.1 LOADING EXTERNAL *DEX**

In the previous chapters, we introduced *ClassLoader*. In Android 6.0 or higher, we need to request permission to read and write onto the SDCard.

Loading an external *dex/apk* file is a combination of these two technologies. The flow is as follows:

1. Download the plug-in *apk* from the server to the mobile's SDCard. That's why we need to request the permissions to read and write onto the SDCard.

2. Read the *dex* of the plug-in app to generate the corresponding *DexClassLoader.*

3. Use the method *loadClass()* of *DexClassLoader* to read classes in this *dex.*

---

* Sample code: https://github.com/Baobaojianqiang/Dynamic0

At the beginning of studying plug-in techniques, I used to upload the plug-in *apk* from my computer to the SDCard and copy it from the SDCard to a specified directory manually. So the app can read the plug-in in this directory.

This solution is quite troublesome. We must do it again and again if the plug-in is frequently modified.

Later, I put the plug-in *apk* into the folder *Assets* in the app, as shown in Figure 6.1. After the app is launched, I write code to copy the plug-ins in the folder *Assets* to the memory. In this way, we simulate the process of downloading plug-ins from the server to the Android device, which is relatively simple and convenient for debugging.

FIGURE 6.1   Put the plug-in in the folder assets of the HostApp.

app-debug.apk is a plug-in. The code of this plug-in is in Dynamic/Plugin1. The code in Plugin1 is very simple, including only one file, Bean. java, shown as follows:

```
package jianqiang.com.plugin1;

public class Bean {
  private String name = "jianqiang";
```

```java
public String getName() {
  return name;
}

public void setName(String paramString) {
  this.name = paramString;
}
}
```

We compile and package the Plugin1 project, generate *app-debug.apk*, and copy this *apk* file into the folder *Assets* in the project Host.

The next step is to load *app-debug.apk* in the Host project. The process has two steps:

1. Copy *app-debug.apk* from the folder *Assets* into the directory */data/ data/files*. We encapsulate this logic into the method *extractAssets()* of *Utils*, and then call this method when the app is launched. In this example, we rewrite the method *attachBaseContext()* of *MainActivity* and invoke the method *extractAssets()* of *Utils*.

```java
@Override
protected void attachBaseContext(Context newBase) {
  super.attachBaseContext(newBase);
  try {
    Utils.extractAssets(newBase, apkName);
  } catch (Throwable e) {
    e.printStackTrace();
  }
}
```

2. Load *dex* into *app-debug.apk*

```java
File extractFile = this.getFileStreamPath(apkName);
dexpath = extractFile.getPath();

fileRelease = getDir("dex", 0); //0 表示Context.
MODE_PRIVATE

classLoader = new DexClassLoader(dexpath,
    fileRelease.getAbsolutePath(), null,
    getClassLoader());
```

From these four code lines, we can get an instance of *ClassLoader*, and can load any class in the *dex* of Plugin1, for example, the class *Bean*.

```
Class mLoadClassBean;
try {
  mLoadClassBean = classLoader.loadClass("jianqiang.
  com.plugin1.Bean");
  Object beanObject = mLoadClassBean.newInstance();

  Method getNameMethod = mLoadClassBean.
  getMethod("getName");
  getNameMethod.setAccessible(true);
  String name = (String) getNameMethod.
  invoke(beanObject);
} catch (Exception e) {
  Log.e("DEMO", "msg:" + e.getMessage());
}
```

Although we have the class *Bean*, we can't use it directly in programming, because class *Bean* doesn't exist in the Host project. We can only use *Bean* using the refection syntax, but as we discussed in Chapter 3, the refection syntax is ugly and difficult to read.

To solve this problem, we introduce another technique, interface-oriented programming.

6.2 INTERFACE-ORIENTED PROGRAMMING*

There are five design principles in the design pattern. Put together the first letters of each design principle, and we get a single word SOLID. The fourth character is I, which represents the principle of dependency inversion, which is defined as follows:

> *Oriented toward interfaces or abstract programming, not specific or implementation-oriented programming.*

Now let's understand this design principle through an example.

First, we create a class library named *MyPluginLibrary*; we set the project HostApp and Plugin1 to rely on *MyPluginLibrary*.

---

* Sample code: https://github.com/Baobaojianqiang/Dynamic1.0

It's easy to create project dependencies in Android Studio. Take the HostApp as an example; refer to Figure 6.2.

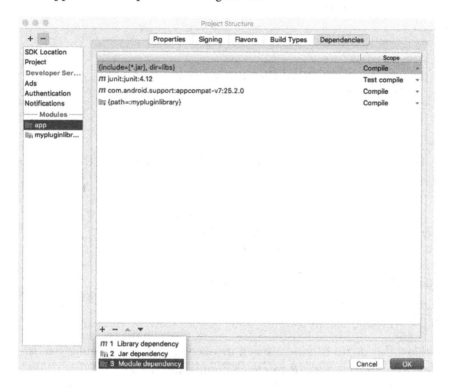

FIGURE 6.2   Add dependencies between projects.

Second, we create an interface *IBean* in *MyPluginLibrary*:

```
public interface IBean {
  String getName();

  void setName(String paramString);
}
```

Third, in the project Plugin1, the class *Bean* implements the interface *IBean*:

```
import com.example.jianqiang.mypluginlibrary.IBean;

public class Bean implements IBean {
  private String name = "jianqiang";
```

```
@Override
public String getName() {
  return name;
}

@Override
public void setName(String paramString) {
  this.name = paramString;
}
}
```

Fourth, in the project HostApp, we write code based on the interface *IBeans*, as follows:

```
Class mLoadClassBean = classLoader.
loadClass("jianqiang.com.plugin1.Bean");
Object beanObject = mLoadClassBean.newInstance();

IBean bean = (IBean) beanObject;
bean.setName("Hello");
tv.setText(bean.getName());
```

Now this demo is done, we can invoke the method *getName()* or *setName()* of the interface *IBeans* to communicate with Plugin1.

Sometimes, we want Plugin1 to push data to the HostApp by itself. We use a callback to satisfy this requirement.

First, in the project *MyPluginLibrary*, we add a method *register()* in the interface *IBean*.

Second, in the project Plugin1, we implement the method *register()* in *Bean*, which will invoke the method *clickButton()*:

```
public class Bean implements IBean {
  ...
  private ICallback callback;

  @Override
  public void register(ICallback callback) {
    this.callback = callback;

    clickButton();
  }
```

```
  public void clickButton() {
    callback.sendResult("Hello: " + this.name);
  }
}
```

Third, in the project HostApp, we invoke the method *register()* of *Bean*, and pass an *ICallback* object as a parameter into the method *register()*:

```
Class mLoadClassBean = classLoader.
loadClass("jianqiang.com.plugin1.Bean");
Object beanObject = mLoadClassBean.newInstance();

IBean bean = (IBean) beanObject;

ICallback callback = new ICallback() {
    @Override
    public void sendResult(String result) {
        tv.setText(result);
    }
};
bean.register(callback);
```

These two demos show the beauty of Interface-Oriented programming.

Up until now, our plug-in app doesn't have a UI. UI is a complex topic in Android, especially in plug-in programming, and we shall talk about this topic in the following chapters.

6.3 PLUG-IN THINNING*

If we find code or logic that can be used in many places, we usually encapsulate it into a class or a library. It's also suitable for plug-in techniques.

Although we encapsulate the logic and code in *MyPluginLibrary*, we compile *MyPluginLibrary* as a *war/jar* file, and all the plug-in apps are dependent on this *war/jar* file and compile this file into the plug-in app. Because all the plug-ins will be packaged into one final app, the size of this final app will be larger.

It's not a good solution. We must take action to make the final app smaller. Let's have a look at the *plugin1.apk* via JadxGUI (Figure 6.3).

We find *MyPluginLibrary* exists in *plugin1.apk*. We want to remove this package in the plug-in app.

* Sample Code: https://github.com/Baobaojianqiang/Dynamic1.1

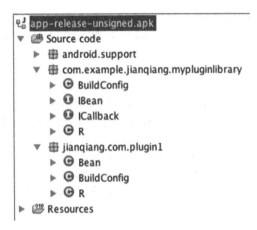

FIGURE 6.3 *plugin1.apk* in JadxGUI.

To solve this problem, we use the keyword "provided," rather than "compile."

In general, we use "compile." Take the Gradle file of the project *Plugin1* as an example:

```
dependencies {
  compile fileTree(dir: "libs", include: ["*.jar"])
  testCompile "junit:junit:4.12"
  compile "com.android.support:appcompat-v7:25.2.0"
  compile project(path: ':MyPluginLibrary')
}
```

Wherever the keyword "compile" appears, the corresponding *jar* file will be packaged into the current *apk*.

We can replace the keyword "compile" with "provided." The keyword "provided" means that the corresponding *jar* package will be used only at "compile" time but won't package into the current *apk*.

The keyword "provided" only supports *jar* packages, not including the module. Therefore, the *MyPluginLibrary* project must be packaged as a *jar* file first. Let's write a task named *makeJar* in *build.gradle* of *MyPluginLibrary*:

```
task clearJar(type: Delete) {
  delete "build/outputs/mypluginlibrary.jar"
}
```

```
task makeJar(type: Copy) {
  from("build/intermediates/bundles/default/")
  into("build/outputs/")
  include("classes.jar")
  rename ("classes.jar", "mypluginlibrary.jar")
}


makeJar.dependsOn(clearJar, build)
```

Click the menu "Sync Project with Gradle files" in Android Studio; we can see a new command named *makeJar* in the Gradle panel (Figure 6.4).

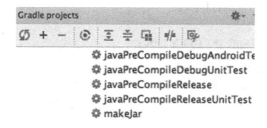

FIGURE 6.4 *makeJar* in Android Studio.

Click the command *makeJar*; we find that a new *jar* named *mypluginlibrary.jar* was generated in the directory *MyPluginLibrary/build/outputs/*.

Copy this jar file to the folder *Lib* of the project Plugin1, and then use the keyword "provided" to reference this *jar* file in *build.gradle*:

```
dependencies {
  compile fileTree(dir: "libs", include: ["*.jar"])
  testCompile "junit:junit:4.12"
  compile "com.android.support:appcompat-v7:25.2.0"
  provided files("lib/classes.jar")
}
```

Note: do not put the *jar* file in the directory *Libs*, it will cause the keyword "provided" not to work.

After we configure the keyword "provided," we package the project Plugin1 again and then use the JadxGUI to see the contents of *plugin1.apk*, and we will now find the package *MyPluginLibrary* doesn't exist, as shown in Figure 6.5.

FIGURE 6.5   Structure of *plugin1.apk* using the keyword provided. Note: The syntax of Gradle 3.0 has replaced the keyword provided with *compileOnly*, and the functionality has not changed.

6.4 DEBUGGING IN PLUG-INS*

Thanks to Android Studio, we can debug plug-ins easily.

We use the code example Dynamic1.1 in Section 6.3, to show how to debug a plug-in app step by step. We put the HostApp, Plugin1, and *MyPluginLibrary* into one Android project so that we can debug from the HostApp to Plugin1 and *MyPluginLibrary*.

1) Create an Android app project named Dynamic1.2. The directory is shown in Figure 6.6.

FIGURE 6.6   The first version of Dynamic1.2.

---

* Code sample: https://github.com/Baobaojianqiang/Dynamic1.2

2) Delete the subdirectory *App* in the directory Dynamic1.2.

3) In the directory Dynamic 1.2, let's create three subdirectories named "HostApp," "Plugin1," and "MyPluginLibrary."

4) Copy and paste from one place to the other in the following table:

From	To
Dynamic1.1/Host/app	Dynamic1.2/ Host
Dynamic1.1/Plugin1/app	Dynamic1.2/ Plugin1
Dynamic1.1/MyPluginLibrary	Dynamic1.2/MyPluginLibrary

5) Modify the file *settings.gradle* in Dynamic1.2:

```
include ":HostApp", ":Plugin1", ":MyPluginLibrary"
```

6) Re-open the project Dynamic1.2. Click the menu "Sync Project with Gradle files" in Android Studio, and the project is shown in Figure 6.7.

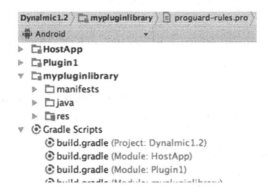

FIGURE 6.7   New directories after the project sync.

Insert the following code in the file *build.gradle* of Plugin1, it helps us to rename the *apk* as "*plugin1.apk*" rather than "*app-debug.apk*" after we compile and build an *apk*, and copy this plug-in app to the folder *Assets* of the HostApp.

```
assemble.doLast {
  android.applicationVariants.all { variant ->
    // Copy Release artifact to HostApp's assets and
    rename
    if (variant.name == "release") {
      variant.outputs.each { output ->
        File originFile = output.outputFile
        println originFile.absolutePath
        copy {
          from originFile
          into "$rootDir/HostApp/src/main/assets"
          rename(originFile.name, "plugin1.apk")
        }
      }
    }
  }
}
```

7) Let's set a breakpoint in the code of project Plugin1, and when we begin to debug this project we'll find that it stops at this breakpoint. Android Studio is powerful for debugging.

6.5 *APPLICATION* PLUG-IN SOLUTIONS*

In the plug-in app, we can also define a custom *Application*, and plug-in apps will do some initial work in the method *onCreate()* of this custom *Application*.

The method *onCreate()* of the custom *Application* in the plug-in app doesn't have a lifecycle. It is only a normal class; it has no chances of being executed. We must invoke it in the method *onCreate()* of the custom *Application* in the

The code examples in this section are based on ZeusStudy1.4; the major changes are located in the method *onCreate()* of *MyApplication*.

```
public class MyApplication extends Application {
  //ignore some code

  @Override
  public void onCreate() {
    super.onCreate();
```

---

* Code sample: https://github.com/Baobaojianqiang/ ZeusStudy1.8

```
for(PluginItem pluginItem: PluginManager.plugins) {

  try {
    Class clazz = PluginManager.mNowClassLoader.
    loadClass(pluginItem.applicationName);
    Application application = (Application)clazz.
    newInstance();

    if(application == null)
      continue;

    application.onCreate();

  } catch (ClassNotFoundException e) {
    e.printStackTrace();
  } catch (InstantiationException e) {
    e.printStackTrace();
  } catch (IllegalAccessException e) {
    e.printStackTrace();
  }
  }
 }
}
```

In addition, we need to provide a method to parse the name of the custom *Application* in the plug-in, as follows:

```
public static String loadApplication(Context context,
File apkFile) {
    Object packageParser = RefInvoke.
    createObject("android.content.pm.PackageParser");
    Class[] p1 = {File.class, int.class};
    Object[] v1 = {apkFile, PackageManager.
    GET_RECEIVERS};
    Object packageObj = RefInvoke.invokeInstanceMethod
    (packageParser, "parsePackage", p1, v1);

    Object obj = RefInvoke.getFieldObject(packageObj,
    "applicationInfo");
    ApplicationInfo applicationInfo =
    (ApplicationInfo)obj;
    return applicationInfo.className;
  }
```

6.6 SUMMARY

This chapter introduces the basic knowledge of plug-in techniques. The "protagonist" in plug-in techniques is *Activity* and *Resource*. We'll talk about these techniques one by one in the following chapters.

CHAPTER 7

Resources in Plug-In

*A*CTIVITY AND *R*ESOURCE ARE twin brothers. If we want to resolve *Activity* in the plug-in completely, we must deal with *Resource* first.

In this chapter, we discuss how to load the resources in the Android system, and then study how to load resources into the plug-in. Finally, we introduce two solutions for changing skin in the app dynamically.

7.1 HOW TO LOAD RESOURCES IN ANDROID

7.1.1 Types of *Resources*

Generally, there are two categories of *Resources* in Android.

One is in the folder *Res* which will be compiled into a binary file. During the compiling process, the file *R.java* will be created; it contains a *HEX* value which corresponds to each *Resource* file, shown as follows:

```
public final class R {
    public static final class anim {
        public static final int abc_fade_in=0x7f050000;
        public static final int abc_fade_out=0x7f050001;
        public static final int abc_grow_fade_in_from_bottom=0x7f050002;
        public static final int abc_popup_enter=0x7f050003;
        public static final int abc_popup_exit=0x7f050004;
        public static final int abc_shrink_fade_out_from_bottom=0x7f050005;
        public static final int abc_slide_in_bottom=0x7f050006;
        public static final int abc_slide_in_top=0x7f050007;
        public static final int abc_slide_out_bottom=0x7f050008;
        public static final int abc_slide_out_top=0x7f050009;
    }
}
```

```
public static final class id {
    public static final int action0=0x7f0b006d;
    public static final int action_bar=0x7f0b0047;
    public static final int action_bar_activity_content=0x7f0b0000;
    public static final int action_bar_container=0x7f0b0046;
    public static final int action_bar_root=0x7f0b0042;
    public static final int action_bar_spinner=0x7f0b0001;
    public static final int action_bar_subtitle=0x7f0b0025;
    public static final int action_bar_title=0x7f0b0024;
    public static final int action_container=0x7f0b006a;
    public static final int action_context_bar=0x7f0b0048;
    public static final int action_divider=0x7f0b0071;
    public static final int action_image=0x7f0b006b;
```

It is easy to access these *Resources*; for example, *R.id.action0*. First, we get an instance of the class *Resources* via the method *getResources()* of *Context*, and then we can fetch any resources through the method *getString()* of this instance, shown as follows:

```
Resources resources = getResources();
String appName = resources.getString(R.string.
app_name)
```

The other category of *Resources* is stored in the folder *Assets*; the files in this folder won't be compiled into a binary file: they are packaged into the *apk* file, so we cannot access them through *R.java*.

The only way to fetch the *Resources* in *Assets* is to use the method *open()* of the *AssetManager*. We can get the instance of *AssetManager* from the method *getAssets()* of the *Resources*. The code is as follows:

```
Resources resources = getResources();
AssetManager am = getResources().getAssets();
InputStream is = getResources().getAssets().
open("filename");
```

7.1.2 *Resources* and *AssetManager*

Resources are like the salesman in an IT company; *AssetManager* is like the programmer. The salesman always faces the customers; the programmer works inside the company.

We can see that *Resource* provides a lot of methods like *getString()*, *getText()*, *getDrawable()*, and so on. In fact, all the methods above invoke the corresponding *private* method of *AssetManager* indirectly. *AssetManager* has no logic and is only a wrapper of *Resources*.

It is not "fair" for *AssetManager*, as it does lots of work, but it is not well known to us, *AssetManager* has only two *public* methods. For example, the method *open()* is used to access the *Resources* in the folder *Assets*.

AssetManager has a method *addAssetPath(String path)*, and we can pass the path of the *apk* to the parameter *path*, then AssetManager and *Resources* can access all the resources in this *apk*.

addAssetPath(String path) is not a *public* method; we can use reflection syntax, put the path of the plug-in *apk* into this method, then the resources of this plug-in *apk* will be thrown into a resource pool. The resources of the current app are already in this pool.

How many times the method *addAssetPath()* is executed depends on the number of plug-in apps (Figure 7.1).

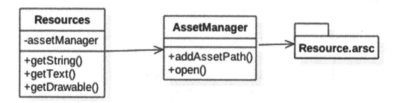

FIGURE 7.1   *Resources* and *AssetManager*.

AssetManager has an internal method to access resources. During the process of the *apk*'s packaging, each resource generates a *HEX* value in *R.java*. But when the app is running, how do we know the mapping between this *HEX* value and the corresponding resource file?

During the process of the *apk*'s packaging, a file named *resources.arsc* will be generated at the same time, which is a *HashTable* that stores all the mapping between the *HEX* value and the corresponding resource file.

7.2 PLUG-IN SOLUTIONS OF *RESOURCES**

Let's try to read a *String* resource from a plug-in app.

1) The code in Plugin1:

```
public class Dynamic implements IDynamic {

@Override
public String getStringForResId(Context context) {
```

---

* https://github.com/BaoBaoJianqiang/dynamic1.3

```
    return context.getResources().getString(R.string.
    myplugin1_hello_world);
  }
}
```

In Plugin1, there is a file *strings.xml* in the folder *Res/Values*, we can find a configuration for this resource:

```
<resources>
  <string name=" myplugin1_hello_world">Hello World</
  string>
</resources>
```

2) The code in the HostApp:

```
public class MainActivity extends AppCompatActivity {

  private AssetManager mAssetManager;
  private Resources mResources;
  private Resources.Theme mTheme;
  private String dexpath = null; //apk file path
  private File fileRelease = null; //decompress folder
  private DexClassLoader classLoader = null;

  private String apkName = "plugin1.apk"; //apk file
  name

  TextView tv;

  @Override
  protected void attachBaseContext(Context newBase) {
    super.attachBaseContext(newBase);
    try {
      Utils.extractAssets(newBase, apkName);
    } catch (Throwable e) {
      e.printStackTrace();
    }
  }

  @SuppressLint("NewApi")
  @Override
  protected void onCreate(Bundle savedInstanceState) {
```

```
   super.onCreate(savedInstanceState);
   setContentView(R.layout.activity_main);
File extractFile = this.getFileStreamPath(apkName);
   dexpath = extractFile.getPath();

   fileRelease = getDir("dex", 0); //0 means Context.
   MODE_PRIVATE

   classLoader = new DexClassLoader(dexpath,
       fileRelease.getAbsolutePath(), null,
       getClassLoader());

   Button btn_6 = (Button) findViewById(R.id.btn_6);

   tv = (TextView)findViewById(R.id.tv);

   // The calling of resource files
   btn_6.setOnClickListener(new View.
   OnClickListener() {
     @Override
     public void onClick(View arg0) {
       loadResources();
       Class mLoadClassDynamic = null;

       try {
         mLoadClassDynamic = classLoader.
         loadClass("jianqiang.com.plugin1.Dynamic");
         Object dynamicObject = mLoadClassDynamic.
         newInstance();

         IDynamic dynamic = (IDynamic) dynamicObject;
         String content = dynamic.
         getStringForResId(MainActivity.this);
         tv.setText(content);
         Toast.makeText(getApplicationContext(),
         content + "", Toast.LENGTH_LONG).show();
       } catch (Exception e) {
         Log.e("DEMO", "msg:" + e.getMessage());
       }
     }
   });
 }
```

```
protected void loadResources() {
  try {
    AssetManager assetManager = AssetManager.class.
    newInstance();
    Method addAssetPath = assetManager.getClass().
    getMethod("addAssetPath", String.class);
    addAssetPath.invoke(assetManager, dexpath);
    mAssetManager = assetManager;
  } catch (Exception e) {
    e.printStackTrace();
  }

  mResources = new Resources(mAssetManager, super.
  getResources().getDisplayMetrics(), super.
  getResources().getConfiguration());
  mTheme = mResources.newTheme();
  mTheme.setTo(super.getTheme());
}

@Override
public AssetManager getAssets() {
  if(mAssetManager == null) {
    return super.getAssets();
  }

  return mAssetManager;
}

@Override
public Resources getResources() {
  if(mResources == null) {
    return super.getResources();
  }

  return mResources;
}

@Override
public Resources.Theme getTheme() {
  if(mTheme == null) {
    return super.getTheme();
  }
```

```
    return mTheme;
  }
}
```

The logic of the code above is divided into four parts:

1) The logic in the method *loadResources()*

Create an instance of *AssetManager* using reflection syntax, invoke the method *addAssetPath()* to add the path of the plug-in app into the resource pool.
This resource pool only contains the resources of Plugin1. This means the instance of *AsssetManager* we created earlier only serves for Plugin1.

2) Based on this instance, we create the corresponding *Resources* and *Theme* by overriding the methods *getAssets()*, *getResources()*, and *getTheme()*. The logic of these three methods is almost the same. Let's take the method *getAssets()* as an example, as follows:

```
@Override
public AssetManager getAssets() {
  if(mAssetManager == null) {
    return super.getAssets();
  }

  return mAssetManager;
}
```

In the method *getAssets()* of *Activity* above, the object *mAssetManager* points to the plug-in app as a default. If *mAssetManager* is *null*, it means that it's a normal app without a plug-in, and we invoke *super.getAssets()* to get the resources. As we introduced in Section 2.7, the parent of *Activity* is *ContextImpl*; if *mAssetManager* is not *null*, we use *mAssetManager* to get the resources in the plug-in.

3) Load the plug-in and generate the corresponding *ClassLoader*:

```
File extractFile = this.getFileStreamPath(apkName);
  dexpath = extractFile.getPath();
```

```
fileRelease = getDir("dex", 0); //0 means Context.
MODE_PRIVATE

classLoader = new DexClassLoader(dexpath,
    fileRelease.getAbsolutePath(), null,
    getClassLoader());
```

4) Use reflection syntax to load the class of the plug-in using inter-face-oriented programming. In this sample, we generate an object *dynamicObject*, and invoke its method *getStringForResId()* to visit the resources in the plug-in app.

```
Class mLoadClassDynamic = classLoader.
loadClass("jianqiang.com.plugin1.Dynamic");
Object dynamicObject = mLoadClassDynamic.
newInstance();

IDynamic dynamic = (IDynamic) dynamicObject;
String content = dynamic.
getStringForResId(MainActivity.this);
tv.setText(content);
```

Now, we will find a perfect solution for loading resources into the plug-in app.

But first we find all the logic is in the *MainActivity* of the HostApp, and refactoring is required. Please refer to the project Dynamic2*, where I move some basic logic to *BaseActivity*; for example, the method *getAssets()*, *getResources()* and *getTheme()*.

7.3 SOLUTIONS FOR CHANGING SKINS†

There are a lot of apps, like games or chatrooms, which have a feature to support the changing of skins dynamically. For example, the emoji in a chatroom app can be downloaded and used immediately.

A simple method for this is to compress all the new emoji pictures into a zip file named *skin1.zip*. After the app downloads this zip file, it will unzip it into a folder named *Skin1* and use the new picture as *skin1/a.png* rather than the original picture *a.png*.

* https://github.com/BaoBaoJianqiang/dynamic2
† https://github.com/BaoBaoJianqiang/dynamic3

We can also implement this requirement based on plug-in technology; we can put all the images into the plug-in app and read each of these resources using *R.java*.

Let's finish this function based on the project Dynamic1.2 in Section 7.3.

1) Generate *plugin1.apk*.

In the project Plugin1, we write a class named *UIUtil* with three methods, *getText()*, *getImage()*, and *getLayout()*, which can fetch strings, images, and layouts from *R.java*.

```
public class UIUtil {
  public static String getTextString(Context ctx){
    return ctx.getResources().getString(R.string.
    hello_message);
  }

  public static Drawable getImageDrawable(Context ctx)
{
    return ctx.getResources().getDrawable(R.drawable.
    robert);
  }

  public static View getLayout(Context ctx){
    return LayoutInflater.from(ctx).inflate(R.layout.
    main_activity, null);
  }
}
```

Then, we put some resources into Plugin1:

- Put an image *robert.png* into *Res/Drawable*, as shown in Figure 7.2.

- Add a string value *hello_message* into *strings.xml* in the folder *Res/ Values*, as follows:

FIGURE 7.2 *robert.png* in Plugin1.

```
<string name="hello_message">Hello</string>
```

- Modify the layout file *main_activity.xml* in the folder *Res/Layout*. There are three buttons placed horizontally in *main_activity.xml*.

We compile and package Plugin1 to generate *plugin1.apk*. Then we put this apk file in the folder *Assets* of the HostApp.

After the app is compiled, rename the *apk* file *plugin1.apk* and put it into the *Assets* folder in the HostApp.

2) Generate *plugin2.apk*

We generate *plugin2.apk* in the same way as *plugin1.apk*, but we do make some modifications in the project Plugin2:

- Rotate the image *robert.png* 180 degrees.

- Update the *String* value "hello_message" to "Hi."

- Update the layout *main_activity.xml*, and place the three buttons vertically.

We also put *plugin2.apk* in the folder *Assets* of the HostApp.

3) Work in the HostApp

First, move the common methods to *BaseActivity*, as follows:

- Load *plugin1.apk* and *plugin2.apk* into the folder *Assets*.

- Generate *ClassLoader* for each plug-in.

- Override the methods *getAssets()*, *getResources()*, and *getTheme()*.

- Write methods *loadResources1()* and *loadResources2()* to generate *AssetManager* for each plug-in.

The code is as follows:

```java
public class BaseActivity extends Activity {

  private AssetManager mAssetManager;
  private Resources mResources;
  private Resources.Theme mTheme;
  private String dexpath1 = null;  //apk file path
  private String dexpath2 = null;  //apk file path
  private File fileRelease = null; //decompression
  path

  protected DexClassLoader classLoader1 = null;
  protected DexClassLoader classLoader2 = null;

  TextView tv;

  @Override
  protected void attachBaseContext(Context newBase) {
    super.attachBaseContext(newBase);

    Utils.extractAssets(newBase, "plugin1.apk");
    Utils.extractAssets(newBase, "plugin2.apk");
  }

  @Override
  protected void onCreate(Bundle savedInstanceState) {
    super.onCreate(savedInstanceState);

    fileRelease = getDir("dex", 0);

    File extractFile1 = this.
    getFileStreamPath("plugin1.apk");
    dexpath1 = extractFile1.getPath();

    classLoader1 = new DexClassLoader(dexpath1,
    fileRelease.getAbsolutePath(), null,
    getClassLoader());


    File extractFile2 = this.
    getFileStreamPath("plugin2.apk");
    dexpath2 = extractFile2.getPath();
```

```
  classLoader2 = new DexClassLoader(dexpath2,
  fileRelease.getAbsolutePath(), null,
  getClassLoader());
}

protected void loadResources1() {
  try {
    AssetManager assetManager = AssetManager.class.
    newInstance();
    Method addAssetPath = assetManager.getClass().
    getMethod("addAssetPath", String.class);
    addAssetPath.invoke(assetManager, dexpath1);
    mAssetManager = assetManager;
  } catch (Exception e) {
    e.printStackTrace();
  }
  Resources superRes = super.getResources();
  mResources = new Resources(mAssetManager,
  superRes.getDisplayMetrics(), superRes.
  getConfiguration());
  mTheme = mResources.newTheme();
  mTheme.setTo(super.getTheme());
}

protected void loadResources2() {
  try {
    AssetManager assetManager = AssetManager.class.
    newInstance();
    Method addAssetPath = assetManager.getClass().
    getMethod("addAssetPath", String.class);
    addAssetPath.invoke(assetManager, dexpath2);
    mAssetManager = assetManager;
  } catch (Exception e) {
    e.printStackTrace();
  }
  Resources superRes = super.getResources();
  mResources = new Resources(mAssetManager,
  superRes.getDisplayMetrics(), superRes.
  getConfiguration());
  mTheme = mResources.newTheme();
  mTheme.setTo(super.getTheme());
}
```

```
@Override
public AssetManager getAssets() {
  return mAssetManager == null ? super.getAssets() :
  mAssetManager;
}

@Override
public Resources getResources() {
  return mResources == null ? super.getResources() :
  mResources;
}

@Override
public Resources.Theme getTheme() {
  return mTheme == null ? super.getTheme() : mTheme;
}
}
```

Second, write an *Activity* named *ResourceActivity* that inherits from *BaseActivity*. In *ResourceActivity*, we click the button "Theme1" to load the skin of *plugin1.apk* and click the button "Theme2" to load the skin of *plugin2.apk*:

```
public class ResourceActivity extends BaseActivity {

  /**
    * List 3 examples : TextView,ImageView,LinearLayout
    */
  private TextView textV;
  private ImageView imgV;
  private LinearLayout layout;

  @Override
  protected void onCreate(Bundle savedInstanceState) {
    super.onCreate(savedInstanceState);
    setContentView(R.layout.activity_resource);

    textV = (TextView) findViewById(R.id.text);
    imgV = (ImageView) findViewById(R.id.imageview);
    layout = (LinearLayout) findViewById(R.id.layout);

    findViewById(R.id.btn1).setOnClickListener(new
    OnClickListener() {
```

```
    @Override
    public void onClick(View arg0) {
      loadResources1();

      doSomething1();
    }
  });

  findViewById(R.id.btn2).setOnClickListener(new
  OnClickListener() {
    @Override
    public void onClick(View v) {
      loadResources2();

      doSomething2();
    }
  });
}

private void doSomething1() {
  try {
    Class clazz = classLoader1.loadClass("jianqiang.
    com.plugin1.UIUtil");

    String str = (String) RefInvoke.
    invokeStaticMethod(clazz, "getTextString",
    Context.class, this);
    textV.setText(str);

    Drawable drawable = (Drawable) RefInvoke.
    invokeStaticMethod(clazz, "getImageDrawable",
    Context.class, this);
    imgV.setBackground(drawable);

    layout.removeAllViews();
    View view = (View) RefInvoke.
    invokeStaticMethod(clazz, "getLayout", Context.
    class, this);
    layout.addView(view);

  } catch (Exception e) {
    Log.e("DEMO", "msg:" + e.getMessage());
  }
}
```

```
private void doSomething2() {
  try {
    Class clazz = classLoader2.loadClass("jianqiang.
    com.plugin1.UIUtil");

    String str = (String) RefInvoke.
    invokeStaticMethod(clazz, "getTextString",
    Context.class, this);
    textV.setText(str);

    Drawable drawable = (Drawable) RefInvoke.
    invokeStaticMethod(clazz, "getImageDrawable",
    Context.class, this);
    imgV.setBackground(drawable);

    layout.removeAllViews();
    View view = (View) RefInvoke.
    invokeStaticMethod(clazz, "getLayout", Context.
    class, this);
    layout.addView(view);

  } catch (Exception e) {
    Log.e("DEMO", "msg:" + e.getMessage());
  }
 }
}
```

Now, let's run the HostApp: we click the button "Theme"1 to see the skin shown in Figure 7.3, and click the button "Theme2" to see the skin shown in Figure 7.4.

FIGURE 7.3 Effect after click "Button1."

FIGURE 7.4   Effect after click "Button2."

There is too much duplicate code in the HostApp. It is necessary to put these plug-ins into a *HashMap* in the *BaseActivity*, shown as follows:

```
public class BaseActivity extends Activity {

  private AssetManager mAssetManager;
  private Resources mResources;
  private Resources.Theme mTheme;

  protected HashMap<String, PluginInfo> plugins = new
  HashMap<String, PluginInfo>();

  @Override
  protected void attachBaseContext(Context newBase) {
    super.attachBaseContext(newBase);

    Utils.extractAssets(newBase, "plugin1.apk");
    Utils.extractAssets(newBase, "plugin2.apk");
  }

  @Override
  protected void onCreate(Bundle savedInstanceState) {
```

```
  super.onCreate(savedInstanceState);

  genegatePluginInfo("plugin1.apk");
  genegatePluginInfo("plugin2.apk");
}

protected void genegatePluginInfo(String pluginName) {
  File extractFile = this.
  getFileStreamPath(pluginName);
  File fileRelease = getDir("dex", 0);
  String dexpath = extractFile.getPath();
  DexClassLoader classLoader = new
  DexClassLoader(dexpath, fileRelease.
  getAbsolutePath(), null, getClassLoader());

  plugins.put(pluginName, new PluginInfo(dexpath,
  classLoader));
}

protected void loadResources(String dexPath) {
  try {
    AssetManager assetManager = AssetManager.class.
    newInstance();
    Method addAssetPath = assetManager.getClass().
    getMethod("addAssetPath", String.class);
    addAssetPath.invoke(assetManager, dexPath);
    mAssetManager = assetManager;
  } catch (Exception e) {
    e.printStackTrace();
  }
  Resources superRes = super.getResources();
  mResources = new Resources(mAssetManager,
  superRes.getDisplayMetrics(), superRes.
  getConfiguration());
  mTheme = mResources.newTheme();
  mTheme.setTo(super.getTheme());
}

@Override
public AssetManager getAssets() {
  return mAssetManager == null ? super.getAssets() :
  mAssetManager;
}
```

```
@Override
public Resources getResources() {
  return mResources == null ? super.getResources() :
  mResources;
}

@Override
public Resources.Theme getTheme() {
  return mTheme == null ? super.getTheme() : mTheme;
}
}
```

Now the code in *ResourceActivity* will be simple:

```
public class ResourceActivity extends BaseActivity {

  /**
   * The widgets that need to replace the theme
   * The example of them: TextView,ImageView,
   LinearLayout
   */
  private TextView textV;
  private ImageView imgV;
  private LinearLayout layout;

  @Override
  protected void onCreate(Bundle savedInstanceState) {
    super.onCreate(savedInstanceState);
    setContentView(R.layout.activity_resource);

    textV = (TextView) findViewById(R.id.text);
    imgV = (ImageView) findViewById(R.id.imageview);
    layout = (LinearLayout) findViewById(R.id.layout);

    findViewById(R.id.btn1).setOnClickListener(new
    OnClickListener() {
      @Override
      public void onClick(View arg0) {
        PluginInfo pluginInfo = plugins.get("plugin1.
        apk");

        loadResources(pluginInfo.getDexPath());
```

```
            doSomething(pluginInfo.getClassLoader());
          }
        });

      findViewById(R.id.btn2).setOnClickListener(new
      OnClickListener() {
        @Override
        public void onClick(View v) {
          PluginInfo pluginInfo = plugins.get("plugin2.
          apk");

          loadResources(pluginInfo.getDexPath());

          doSomething(pluginInfo.getClassLoader());
        }
      });
    }

    private void doSomething(ClassLoader cl) {
      try {
        Class clazz = cl.loadClass("jianqiang.com.
        plugin1.UIUtil");

        String str = (String) RefInvoke.
        invokeStaticMethod(clazz, "getTextString",
        Context.class, this);
        textV.setText(str);

        Drawable drawable = (Drawable) RefInvoke.
        invokeStaticMethod(clazz, "getImageDrawable",
        Context.class, this);
        imgV.setBackground(drawable);

        layout.removeAllViews();
        View view = (View) RefInvoke.
        invokeStaticMethod(clazz, "getLayout", Context.
        class, this);
        layout.addView(view);

      } catch (Exception e) {
        Log.e("DEMO", "msg:" + e.getMessage());
      }
    }
}
```

Changing skins using the plug-in technique is very easy. Plugin1 is a template; we can generate a new skin plug-in easily using Plugin1, without changing code, only to replace the resources in the folder *Res*.

7.4 ANOTHER PLUG-IN SOLUTION FOR CHANGING SKINS*

The sample code in this section is Dynamic3.2; it's based on Dynamic3.1, and we only modify the method *doSomething()* of *ResourceActivity*.

In the example in Section 7.3, we use *R.drawable.robert* to see the resources in the plug-in app.

In fact, in the HostApp, we can directly access the inner class of the plug-in app in *R.java*, such as *jianqiang.com.plugin1.R$string*, and then we can fetch the resources in the plug-in from this inner class, shown as follows:

```
Class stringClass = cl.loadClass("jianqiang.com.
plugin1.R$string");
int resId1 = (int) RefInvoke.getStaticFieldObject
(stringClass, "hello_message");
textV.setText(getResources().getString(resId1));

Class drawableClass = cl.loadClass("jianqiang.com.
plugin1.R$drawable");
int resId2 = (int) RefInvoke.getStaticFieldObject
(drawableClass, "robert");
imgV.setBackground(getResources().
getDrawable(resId2));

Class layoutClazz = cl.loadClass("jianqiang.com.
plugin1.R$layout");
int resId3 = (int) RefInvoke.getStaticFieldObject
(layoutClazz, "main_activity");
View view = (View) LayoutInflater.from(this).
inflate(resId3, null);
layout.removeAllViews();
layout.addView(view);
```

In this solution, we don't need the class *UIUtil, plugin1.apk* because it is only a container storing the resources and *R.java*.

---

* Sample code: https://github.com/BaoBaoJianqiang/dynamic3.2

7.5 SUMMARY

This chapter gives a detailed description of the *Resources* in Android systems. Based on this mechanism, we use a reflection syntax to invoke the method *addAssetPath()* of *AssetManager* to load resources into the plug-in app.

We introduced two solutions for changing skins using plug-in techniques.

CHAPTER 8

The Plug-In Solution of Four Components

THIS CHAPTER INTRODUCES A plug-in solution for the four components in the Android system: *Activity, Service, ContentProvider,* and *BroadcastReceiver.*

8.1 THE SIMPLEST PLUG-IN SOLUTION

This section introduces the simplest solutions provided by plug-ins which are applicable to all the components, including *Activity, Service, ContentProvider,* and *BroadcastReceiver.* These solutions involve the following techniques:

1) Combine all the *dexes* of the plug-ins to load classes defined in the plug-in.

2) Pre-declare *Activity, Service, ContentProvider,* and *BroadcastReceiver* of the plug-in in the HostApp's *AndroidManifest.xml.* Of course, if there are hundreds of *Activities* in the plug-in, this is not a good solution.

3) Merge all the resources of the plug-in into a resources pool. However, this solution may result in the conflict of resource IDs from different plug-ins.

8.1.1 Pre-Declare *Activity* and *Service* of the Plug-In in the HostApp's *AndroidManifest.xml**

As we mentioned earlier, all components including *Activity, Service, ContentProvider,* and *BroadcastReceiver* in a plug-in are only normal classes; the Android system doesn't recognize them as components at all.

To enable the HostApp to recognize them, we must declare them in the HostApp's *AndroidManifest.xml*. This is the simplest plug-in solution, because we needn't hook a class.

Look at the sample ZeusStudy1.0, as shown in Figure 8.1. Plugin1 has a service named *TestService1*.

Correspondingly, in the *AndroidManifest.xml* of the HostApp, we declare *TestService1* as follows:

```
<service android:name="jianqiang.com.plugin1.
TestService1" />
```

FIGURE 8.1   Project structure of Plugin1.

___

* Sample code: https://github.com/BaoBaoJianqiang/ZeusStudy1.0

8.1.2 Combine the *Dex**

In the HostApp, how do you load the classes from the plug-in? There are three ways:

1) Use *ClassLoader* to load the classes into the plug-in; it's not convenient when we switch *ClassLoader* to load the classes from different plug-ins. We previously introduced this in Section 6.1.

2) Combine all the *dex* files of the HostApp and the plug-ins, and use an array *dexElements* to store all the *dex* files. Once all the *dex* files of the plug-ins are merged into the array *dexElements*, the *dex* of the HostApp is also in this array; we can load any class whether it is into the HostApp or the plug-in; the implementation is as follows:

```
public final class BaseDexClassLoaderHookHelper {

public static void patchClassLoader(ClassLoader cl,
File apkFile, File optDexFile)
    throws IllegalAccessException,
    NoSuchMethodException, IOException,
    InvocationTargetException,
    InstantiationException, NoSuchFieldException {
// Obtain BaseDexClassLoader : pathList
Object pathListObj = RefInvoke.
getFieldObject(DexClassLoader.class.
getSuperclass(), cl, "pathList");

// Obtain PathList: Element[] dexElements
Object[] dexElements = (Object[]) RefInvoke.
getFieldObject(pathListObj, "dexElements");

// Element type
Class<?> elementClass = dexElements.getClass().
getComponentType();

// Create an array to replace the original array
Object[] newElements = (Object[]) Array.
newInstance(elementClass, dexElements.length + 1);
```

---

* Sample code: https://github.com/BaoBaoJianqiang/ZeusStudy1.0

```
// Construct a Plug-In Element(File file, boolean
isDirectory, File zip, DexFile dexFile) This
constructor
Class[] pl = {File.class, boolean.class, File.
class, DexFile.class};
Object[] vl = {apkFile, false, apkFile, DexFile.
loadDex(apkFile.getCanonicalPath(), optDexFile.
getAbsolutePath(), 0)};
Object o = RefInvoke.createObject(elementClass,
pl, vl);

Object[] toAddElementArray = new Object[] { o };
// Copy the original elements
System.arraycopy(dexElements, 0, newElements, 0,
dexElements.length);
// The element of the Plug-In is copied in
System.arraycopy(toAddElementArray, 0,
newElements, dexElements.length,
toAddElementArray.length);

// replace
RefInvoke.setFieldObject(pathListObj,
"dexElements", newElements);
  }
}
```

3) Replace the original *ClassLoader* of Android with a *ClassLoader* written by me. We can add our own logic into this custom *ClassLoader* to iterate each *ClassLoader* of the plug-in. I will introduce this solution in detail in Section 8.2.6.

8.1.3 Start a *Service* of the Plug-In*

Since we have finished the actions in Section 8.1.1 and 8.1.2, we can start a *Service* of the plug-in in the HostApp, as follows:

```
Intent intent = new Intent();
String serviceName = "jianqiang.com.plugin1.
TestService1";
intent.setClassName(this, serviceName);
startService(intent);
```

---

* Sample code: https://github.com/BaoBaoJianqiang/ZeusStudy1.0

8.1.4 *Resources* in *Activity**

In Section 8.1.1, 8.1.2, and 8.1.3, we loaded a *Service* of the plug-in into the HostApp.

All of the other components, *Activity*, *ContentProvider*, and *BroadcastReceiver* of the plug-in can be loaded in the same way.

The plug-in solutions for *Service*, *ContentProvider*, and *BroadcastReceiver* are simple: we only need to merge all the *dex* files of the plug-in into one array and pre-declare them in the *AndroidManifest.xml*.

The solution for *Activity* is a bit complicated.

Activity is heavily dependent on resources. Therefore, we must resolve the problem of how to load the resources into the plug-in.

I introduced the relationship between *AssetManager* and *Resources* in Chapter 7. There is a method *addAssetPath(String path)* of *AssetManager*. We can pass all the paths from the plug-ins into this method; all the paths make up a *String* separated with a comma or a semicolon. Now *AssetManager* becomes a superstar that has all the resources, including the HostApp and the plug-in. We store this *AssetManager* in a global class *PluginManager*; this means we can load any class into the HostApp and the plug-in.

The logic above is implemented as follows (in *MyApplication*):

```
private static void reloadInstalledPluginResources() {
  try {
    AssetManager assetManager = AssetManager.class.
    newInstance();
    Method addAssetPath = AssetManager.class.
    getMethod("addAssetPath", String.class);

    addAssetPath.invoke(assetManager, mBaseContext.
    getPackageResourcePath());
    addAssetPath.invoke(assetManager, pluginItem1.
    pluginPath);


    Resources newResources = new
    Resources(assetManager,
      mBaseContext.getResources().
      getDisplayMetrics(),
```

```
        mBaseContext.getResources().
        getConfiguration());

    RefInvoke.setFieldObject (mBaseContext,
    "mResources", newResources);
    // This is the main need to replace, if you do
    not support the Plug-In runtime update, just
    leave this one
    RefInvoke.setFieldObject (mPackageInfo,
    "mResources", newResources);

    mNowResources = newResources;
    // Need to clean up the mTheme object, otherwise
    it will report an error when loading resources
    through inflate mode
    // If the activity dynamically loads the Plug-In,
    you need to set the activity's mTheme object to
    null
    RefInvoke.setFieldObject (mBaseContext, "mTheme",
    null);
  } catch (Throwable e) {
    e.printStackTrace();
  }
```

Activity in the plug-in must implement the *ZeusBaseActivity*. We override the method *getResources()* in *ZeusBaseActivity*, and it helps us to fetch the resources of the plug-in using *PluginManager*.

```
public class ZeusBaseActivity extends Activity {

  @Override
  public Resources getResources() {
    return PluginManager.mNowResources;
  }
}
```

The code for *TestActivity1* in the plug-in is as follows (it uses the layout *activity_test1.xml* in the plug-in):

```
public class TestActivity1 extends ZeusBaseActivity {
  private final static String TAG = "TestActivity1";
```

```
@Override
protected void onCreate(Bundle savedInstanceState) {
  super.onCreate(savedInstanceState);
  setContentView(R.layout.activity_test1);

  findViewById(R.id.button1).setOnClickListener
  (new View.OnClickListener() {
    @Override
    public void onClick(View v) {
      try {
        Intent intent = new Intent();

        String activityName = "jianqiang.com.hostapp.
        ActivityA";
        intent.setComponent(new ComponentName
        ("jianqiang.com.hostapp", activityName));

        startActivity(intent);

      } catch (Exception e) {
        e.printStackTrace();
      }
    }
  });
}
}
```

Now that the simplest plug-in solution for *Activity* is completed, we can jump from the *Activity* of the plug-in app to the *Activity* of the HostApp.

However, there is a fatal problem: we must pre-declare all the components, such as *Activity, Service, ContentProvider,* and *BroadcastReceiver* of the plug-in into the HostApp's *AndroidManifest.xml;* we can't add a new *Activity* or *Service* after the app has been compiled and packaged.

In fact, the number of *Services, BroadcastReceivers,* or *ContentProviders* used in an app is small, so it makes sense that we pre-define these components of the plug-in in the HostApp. Normally we don't add a new *Service* to the plug-in. If we want to add a new *Service* to the Plug-In, we can wait for the next release of the app.

But *Activity* is different from the other three components. *Activity* is widely used in apps; the number of *Activities* is large. We modify an *Activity* or add a new *Activity* to the plug-in frequently. So, we cannot pre-declare these new Activities in the *AndroidManifest.xml* in the HostApp. We shall resolve this problem in Section 8.2.

8.2 A PLUG-IN SOLUTION FOR ACTIVITY

Activity is widely used in Android apps.

There are three problems with the plug-in solutions for *Activity*:

1) In the HostApp, how can we load the class of a plug-in app?

2) In the HostApp, how can we load the resources of a plug-in app?

3) In the HostApp, how can we load the *Activity* of a plug-in app?

This section focuses on these three problems.

8.2.1 Launch an *Activity* of a Plug-In Not Declared in the *AndroidManifest.xml* of the HostApp*

We introduced a solution in Section 5.4 on to how to launch an *Activity* that is not declared in the *AndroidManifest.xml*.

This solution is also suitable for *Activity* in plug-ins. We pre-declare a *StubActivity* in the HostApp, and when we want to launch an *Activity* from a plug-in we replace it with *StubActivity*.

In the following example, we use *AMSHookHelper* to accomplish this goal.

The method *hookAMN()* of *AMSHookHelper* replaces the *Activity* of the plug-in with a *StubActivity* pre-declared in the *AndroidManifest.xml* of the HostApp. The logic of replacement occurs in *MockClass1* by hooking *AMN*.

The method *hookActivityThread()* of *AMSHookHelper* replaces *StubActivity* with the original *Activity* of the plug-in which is the one to be launched. The logic of replacement occurs in *MockClass2* by hooking *ActivityThread*.

The code of *AMSHookHelper* is as follows:

```
public class AMSHookHelper {

  public static final String EXTRA_TARGET_INTENT =
  "extra_target_intent";
```

---

* Sample code: https://github.com/BaoBaoJianqiang/ActivityHook1

```java
/**
 * Hook AMS
 * The main operation is to temporarily replace the
 plug-in Activity that is actually needed to start
 the StubActivity declared in the AndroidManifest.
 xml, and then cheat the AMS
 */
public static void hookAMN() throws
ClassNotFoundException,
    NoSuchMethodException, InvocationTargetException,
    IllegalAccessException, NoSuchFieldException {

  //Get the gDefault singleton of AMN, gDefault is
  final and static
  Object gDefault = RefInvoke.
  getStaticFieldObject("android.app.
  ActivityManagerNative", "gDefault");

  // gDefault is an instance of android.util.Singleton<T>.
  We get the mInstance field from this singleton
  Object mInstance = RefInvoke.
  getFieldObject("android.util.Singleton", gDefault,
  "mInstance");

  // Create a proxy instance of MockClass1 and replace
  this field. Let the proxy object deal with it.
  Class<?> classB2Interface = Class.
  forName("android.app.IActivityManager");
  Object proxy = Proxy.newProxyInstance(
      Thread.currentThread().getContextClassLoader(),
      new Class<?>[] { classB2Interface },
      new MockClass1(mInstance));

  //Replace the mInstance field of mDefault with
  proxy object
  RefInvoke.setFieldObject("android.util.Singleton",
  gDefault, "mInstance", proxy);
}

/**
 * Because we cheat AMS with a StubActivity, we have to
 replace back the Activity we really need to start now.
 * Otherwise it will really start a StubActivity.
```

```
 * To eventually start the Activity, an inner class
 of ActivityThread called H will do the work
 * H will forward the message through it's callback
 */
public static void hookActivityThread() throws
Exception {

  // Get the current ActivityThread object firstly
  Object currentActivityThread = RefInvoke.
  getStaticFieldObject("android.app.ActivityThread",
  "sCurrentActivityThread");

  // Since ActivityThread has only one process, we
  get the mH of this object
  Handler mH = (Handler) RefInvoke.getFieldObject(cu
  rrentActivityThread, "mH");

  //Replace mCallback field of Handler with new
  MockClass2(mH)
  RefInvoke.setFieldObject(Handler.class,
      mH, "mCallback", new MockClass2(mH));
  }
}
```

The code of *MockClass1* is as follows:

```
class MockClass1 implements InvocationHandler {

  private static final String TAG = "MockClass1";

  Object mBase;

  public MockClass1(Object base) {
    mBase = base;
  }

  @Override
  public Object invoke(Object proxy, Method method,
  Object[] args) throws Throwable {

    Log.e("bao", method.getName());

    if ("startActivity".equals(method.getName())) {
```

```java
// Only intercept this method
// Replace parameters as you want and even can replace
the original Activity to start another Activity.

// Find the first Intent object in the parameter
Intent raw;
int index = 0;

for (int i = 0; i < args.length; i++) {
  if (args[i] instanceof Intent) {
    index = i;
    break;
  }
}
raw = (Intent) args[index];

Intent newIntent = new Intent();

// The packageName of StubActivity
String stubPackage = "jianqiang.com.
activityhook1";

// Replace the Plug-In Activity that is actually
needed to start the StubActivity temporarily
ComponentName componentName = new
ComponentName(stubPackage, StubActivity.class.
getName());
newIntent.setComponent(componentName);

// Save TargetActivity
newIntent.putExtra(AMSHookHelper.EXTRA_TARGET_
INTENT, raw);

// Replace Intent to achieve the purpose of
deceiving AMS
args[index] = newIntent;

Log.d(TAG, "hook success");
return method.invoke(mBase, args);
}

return method.invoke(mBase, args);
}
}
```

The code of *MockClass2* is as follows:

```
class MockClass2 implements Handler.Callback {

  Handler mBase;

  public MockClass2(Handler base) {
    mBase = base;
  }

  @Override
  public boolean handleMessage(Message msg) {

    switch (msg.what) {
      // The value of "LAUNCH_ACTIVITY" of
      ActivityThread is 100
      // Use reflection is the best way, we use hard
      coding here for simplicity
      case 100:
        handleLaunchActivity(msg);
        break;
    }

    mBase.handleMessage(msg);
    return true;
  }

  private void handleLaunchActivity(Message msg) {
    // For simplicity, get the TargetActivity
    directly;

    Object obj = msg.obj;

    // Restore the target intent
    Intent raw = (Intent) RefInvoke.
    getFieldObject(obj, "intent");

    Intent target = raw.getParcelableExtra(AMSHookHelp
    er.EXTRA_TARGET_INTENT);
    raw.setComponent(target.getComponent());
  }
}
```

8.2.2 Solution 1: Based on *Dynamic-Proxy**

There are many plug-in solutions for *Activity*, which are roughly divided into two categories:

- The solution based on *Dynamic-Proxy*, represented by *DroidPlugin*[†] written by Yong Zhang. Which loads *Activity*, *Service*, *BroadcastReceiver*, and *ContentProvider* from the plug-in app through hooking the internal API in the Android system.

- The solution based on *Static-Proxy*, represented by *DL*[‡] written by Yugang Ren. Which loads all *Activities* in the plug-in through *ProxyActivity*.

This chapter focuses on plug-in solutions based on *Dynamic-Proxy* to support *Activity*.

8.2.2.1 The Process of Launching an Activity

In the last step in the process of launching an Activity, the communication between *ActivityThread* and *H* is the key point, as shown in Figure 8.2.

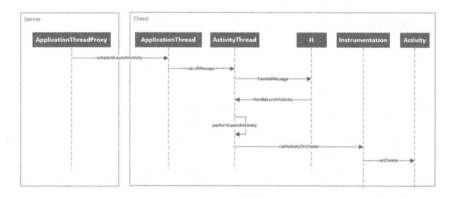

FIGURE 8.2 The last step in the process of launching *Activity*.

---

* Sample code: https://github.com/BaoBaoJianqiang/ActivityHook1
† The GitHub address of DroidPlugin: https://github.com/Qihoo360/DroidPlugin
‡ The GitHub address of DL: https://github.com/singwhatiwanna/dynamic-load-apk

1) ActivityThread invokes the method handleMessage() of H:

```
public void handleMessage(Message msg) {
    switch (msg.what) {
      case LAUNCH_ACTIVITY: {
        final ActivityClientRecord r =
        (ActivityClientRecord) msg.obj;

        r.packageInfo = getPackageInfoNoCheck(
            r.activityInfo.applicationInfo,
            r.compatInfo);
        handleLaunchActivity(r, null);
      } break;
      }
    }
```

r.packageInfo is a type of *LoadedApk*. *LoadedApk* contains information about the current *apk*. We can get information such as *Activity, Service, BroadcastReceiver,* and *ContentProvider*.

The method *getPackageInfoNoCheck()* returns an instance of *LoadedApk*, and it will invoke the method *getPackageInfo()* indirectly:

```
public final LoadedApk getPackageInfoNoCheck(Applica
tionInfo ai,
    CompatibilityInfo compatInfo) {
  return getPackageInfo(ai, compatInfo, null, false,
  true, false);
}

private LoadedApk getPackageInfo(ApplicationInfo
aInfo, CompatibilityInfo compatInfo, ClassLoader
baseLoader, boolean securityViolation, boolean
includeCode, boolean registerPackage) {
  final boolean differentUser = (UserHandle.
  myUserId() != UserHandle.getUserId(aInfo.uid));
  synchronized (mResourcesManager) {
    WeakReference<LoadedApk> ref;
    if (differentUser) {
      ref = null;
    } else if (includeCode) {
      ref = mPackages.get(aInfo.packageName);
    } else {
      ref = mResourcePackages.get(aInfo.packageName);
    }
```

```
LoadedApk packageInfo = ref != null ? ref.get() :
null;
if (packageInfo == null || (packageInfo.
mResources != null
    && !packageInfo.mResources.getAssets().
    isUpToDate())) {
  packageInfo =
    new LoadedApk(this, aInfo, compatInfo,
    baseLoader,
        securityViolation, includeCode &&
        (aInfo.flags&ApplicationInfo.FLAG_HAS_
        CODE) != 0, registerPackage);
  }
  return packageInfo;
 }
}
```

The method *getPackageInfo()* is used to check the cache; for example, there is a code line in this method:

```
ref = mPackages.get(aInfo.packageName);
```

mPackages is a cache object, storing all the instances of *LoadedApk*. If we can't find a suitable object in the cache *mPackages*, we'll create a new *LoadedApk* object and throw it into the cache.

2) Perform the method *LaunchActivity()* of *ActivityThread*:

```
private Activity performLaunchActivity(ActivityClien
tRecord r, Intent customIntent)
  Activity activity = null;

  java.lang.ClassLoader cl = r.packageInfo.
  getClassLoader();
  activity = mInstrumentation.newActivity(
      cl, component.getClassName(), r.intent);
  return activity;
}
```

When we invoke the method *newActivity()* of *mInstrumentation*, we need specify the value of the object *cl*, and we can retrieve *cl* from *r.packageInfo.getClassLoader()*.

r.packageInfo is an instance of *LoadedApk*, and we can get this instance from the cache; if it doesn't exist in the cache, we create a new one.

If it's a normal app without a plug-in, the instance of *cl* is the *ClassLoader* of the HostApp.

If it is a plug-in, the instance of *cl* is the *ClassLoader* of the plug-in.

8.2.2.2 Add a Plug-In Activity to the Cache

We have introduced the mechanisms for launching an *Activity*, now let's find the plug-in solution of *Activity*.

There are two steps:

1) Create an instance of *LoadedApk* and put this instance into the cache. As I introduced in Section 8.2.2.1, *mPackages* is the cache, and the method *getPackageInfo()* will retrieve this instance directly from *mPackages*.

2) Get the field *mClassLoader* of the object *LoadedApk* and set it as the *ClassLoader* of the plug-in *apk*.

The code is as follows:

```
public class LoadedApkClassLoaderHookHelper {

    public static Map<String, Object> sLoadedApk = new
    HashMap<String, Object>();

    public static void hookLoadedApkInActivityThread(F
    ile apkFile) throws ClassNotFoundException,
        NoSuchMethodException, InvocationTargetException,
        IllegalAccessException, NoSuchFieldException,
        InstantiationException {

        // Get the current ActivityThread object firstly
        Object currentActivityThread = RefInvoke.
        invokeStaticMethod("android.app.ActivityThread",
        "currentActivityThread");

        // Get the mPackages field, which caches the dex
        package information
        Map mPackages = (Map) RefInvoke.getFieldObject(cur
        rentActivityThread, "mPackages");
```

```java
// Prepare two parameters
// android.content.res.CompatibilityInfo
Object defaultCompatibilityInfo = RefInvoke.
getStaticFieldObject("android.content.res.
CompatibilityInfo", "DEFAULT_COMPATIBILITY_INFO");
// Get ApplicationInfo information from Apk
ApplicationInfo applicationInfo = generateApplicat
ionInfo(apkFile);

//call getPackageInfoNoCheck method of
ActivityThread, the above two fields are
parameters.
Class[] p1 = {ApplicationInfo.class, Class.
forName("android.content.res.CompatibilityInfo")};
Object[] v1 = {applicationInfo,
defaultCompatibilityInfo};
Object loadedApk = RefInvoke.invokeInstanceMethod(
currentActivityThread, "getPackageInfoNoCheck",
p1, v1);

// Create a Plug-In ClassLoader
String odexPath = Utils.getPluginOptDexDir(applica
tionInfo.packageName).getPath();
String libDir = Utils.
getPluginLibDir(applicationInfo.packageName).
getPath();
ClassLoader classLoader = new
CustomClassLoader(apkFile.getPath(), odexPath,
libDir, ClassLoader.getSystemClassLoader());
RefInvoke.setFieldObject(loadedApk,
"mClassLoader", classLoader);

// Put the Plug-In LoadedApk object into the cache
WeakReference weakReference = new
WeakReference(loadedApk);
mPackages.put(applicationInfo.packageName,
weakReference);

// Because it is a weak reference, we must keep a
copy somewhere, otherwise it is easy to be GC.
sLoadedApk.put(applicationInfo.packageName,
loadedApk);
    }
}
```

We use reflection to execute the method *getPackageInfoNoCheck()* dynamically. This method has two parameters. The type of one parameter is *ApplicationInfo*. It's complex to prepare this parameter. So we write a method *generateApplicationInfo()* to generate the instance of *ApplicationInfo*, shown as follows:

```
public static ApplicationInfo
generateApplicationInfo(File apkFile)
    throws ClassNotFoundException,
    NoSuchMethodException, IllegalAccessException,
    InstantiationException,
    InvocationTargetException, NoSuchFieldException {

    // Find out the core class that needs reflection:
    android.content.pm.PackageParser
    Class<?> packageParserClass = Class.
    forName("android.content.pm.PackageParser");
    Class<?> packageParser$PackageClass = Class.
    forName("android.content.
    pm.PackageParser$Package");
    Class<?> packageUserStateClass = Class.
    forName("android.content.pm.PackageUserState");

    // Get our final goal firstly:
    generateApplicationInfo method
    // API 23 !
    // public static ApplicationInfo
    generateApplicationInfo(Package p, int flags,
    //  PackageUserState state) {
    // Other Android versions do not guarantee this.


    // First, we have to create a Package object for
    this method
    // This object can be returned by the android.
    content.pm.PackageParser#parsePackage method.
    // Create a PackageParser object
    Object packageParser = packageParserClass.
    newInstance();

    // call PackageParser.parsePackage method to parse
    Apk information
```

```
//It's actually an android.content.
pm.PackageParser.Package object
Class[] p1 = {File.class, int.class};
Object[] v1 = {apkFile, 0};
Object packageObj = RefInvoke.invokeInstanceMethod
(packageParser, "parsePackage", p1, v1);


// The third parameter mDefaultPackageUserState
uses the default constructor
Object defaultPackageUserState =
packageUserStateClass.newInstance();

// Ready To Go!
Class[] p2 = {packageParser$PackageClass, int.
class, packageUserStateClass};
Object[] v2 = {packageObj, 0,
defaultPackageUserState};
ApplicationInfo applicationInfo =
(ApplicationInfo)RefInvoke.invokeInstanceMethod(pa
ckageParser, "generateApplicationInfo", p2, v2);

String apkPath = apkFile.getPath();
applicationInfo.sourceDir = apkPath;
applicationInfo.publicSourceDir = apkPath;

return applicationInfo;
}
```

Let's analyze the code above. Invoke the method *generateApplica-tionInfo()* of *PackageParser* using reflection to generate an instance of *ApplicationInfo*. This method is always modified in different visions of Android systems, so the solution in this book is only suitable for Android API 23. The plug-in framework *DroidPlugin* supports all versions of the Android system.

Although we have put the instance of *LoadedApk* into the cache *mPackages*, the key in the cache is still wrong. Let's have a look at a snippet of the source code from the Android system:

```
public void handleMessage(Message msg) {
    switch (msg.what) {
      case LAUNCH_ACTIVITY: {
```

```
      final ActivityClientRecord r =
      (ActivityClientRecord) msg.obj;

      r.packageInfo = getPackageInfoNoCheck(
          r.activityInfo.applicationInfo,
          r.compatInfo);
      handleLaunchActivity(r, null);
    } break;
    }
}

public final LoadedApk getPackageInfoNoCheck(Applica
tionInfo ai,
    CompatibilityInfo compatInfo) {
  return getPackageInfo(ai, compatInfo, null, false,
  true, false);
}

private LoadedApk getPackageInfo(ApplicationInfo
aInfo, CompatibilityInfo compatInfo, ClassLoader
baseLoader, boolean securityViolation, boolean
includeCode, boolean registerPackage) {
    if (differentUser) {
      ref = null;
    } else if (includeCode) {
      ref = mPackages.get(aInfo.packageName);
    } else {
      ref = mResourcePackages.get(aInfo.packageName);
    }
}
```

The method *mPackages.get(aInfo.packageName)* is used to search the cache, which means the key is *aInfo.packageName*, the object *aInfo* comes from *r.activityInfo.applicationInfo*, the object *r* is the field *obj* of the parameter *msg* of the method *handleMessage(Message msg)*.

Because we cheat the *AMS*, the field *packageName* of *msg.obj.activity-Info.applicationInfo* is still *jianqiang.com.activityhook1*, and we need to replace it with the *packageName* of the plug-in, *jianqiang.com.testactivity*. The code is as follows:

```
ActivityInfo activityInfo = (ActivityInfo) RefInvoke.
getFieldObject(obj, "activityInfo");
```

```
activityInfo.applicationInfo.packageName = target.
getPackage() == null?target.getComponent().
getPackageName() : target.getPackage();
```

We put the code above at the end of the method *handleLaunchActivity* of *MockClass2*.

8.2.2.3 Solution 1 of Loading Class in a Plug-In: Create DexClassLoader for Each Plug-In apk

We introduced the method *hookLoadedApkInActivityThread()* of *LoadedApkClassLoaderHookHelper* in Section 8.2.2.2. In this method, we find something interesting, shown as follows:

```
String odexPath = Utils.getPluginOptDexDir(applica
tionInfo.packageName).getPath();
String libDir = Utils.
getPluginLibDir(applicationInfo.packageName).
getPath();
ClassLoader classLoader = new
CustomClassLoader(apkFile.getPath(), odexPath,
libDir, ClassLoader.getSystemClassLoader());
RefInvoke.setFieldObject(loadedApk,
"mClassLoader", classLoader);
```

Each plug-in is an instance of *LoadedApk*. We replace the field *mClassLoader* of *LoadedApk* with a custom *ClassLoader* named *CustomClassLoader*.

The definition of the *CustomClassLoader* is as follows; it is actually a subclass of *DexClassLoader*:

```
public class CustomClassLoader extends DexClassLoader {

  public CustomClassLoader(String dexPath, String
  optimizedDirectory, String libraryPath, ClassLoader
  parent) {
    super(dexPath, optimizedDirectory, libraryPath,
    parent);
  }
}
```

We can launch Activity and every class of the plug-in app by CustomClassLoader. When the app user navigates from the HostApp to the plug-in app, the app will use CustomClassLoader. When the app user leaves the plug-in app and goes back to the HostApp, the app will use its original *ClassLoader.*

8.2.2.4 Hooking More Classes
Implementing plug-in technology via hooking can cause an interesting bug. When we click the button in the *Activity* of the plug-in, it will throw up an exception as follows:

```
Unable to get package info for jianqiang.com.
testactivity; is package not installed?
```

We can search the exception in the source code of the Android system, and find out the problem is in the method *initializeJavaContextClass-Loader* of *LoadedApk*:

```
private void initializeJavaContextClassLoader() {
  IPackageManager pm = ActivityThread.
  getPackageManager();
  android.content.pm.PackageInfo pi;
  try {
    pi = pm.getPackageInfo(mPackageName, 0,
    UserHandle.myUserId());
  } catch (RemoteException e) {
    throw new IllegalStateException("Unable to get
    package info for "
       + mPackageName + "; is system dying?", e);
  }
  if (pi == null) {
    throw new IllegalStateException("Unable to get
    package info for "
       + mPackageName + "; is package not
       installed?");
  }
}
```

The variable *pi* is *null*, so it throws up an exception. Let's have a look at the generation of the *pi*,

```
pi= pm.getPackageInfo();
```

If we replace the object *pm* with a new object *Proxy*, and override the method *getPackageInfo()* of *Proxy*, *pi* will not be *null*;

Let's have a look at how to get the object *pm*, as follows:

```
public final class ActivityThread {
  private static ActivityThread
  sCurrentActivityThread;

  static IPackageManager sPackageManager;

  public static IPackageManager getPackageManager() {
    if (sPackageManager != null) {
      return sPackageManager;
    }
    IBinder b = ServiceManager.getService("package");
    sPackageManager = IPackageManager.Stub.
    asInterface(b);
    return sPackageManager;
  }
}
```

The method *getPackageManager()* of *ActivityThread* returns the object *pm*, and this method actually returns the field *sPackageManager* of the *ActivityThread*, so we can hook this field, as follows:

```
private static void hookPackageManager() throws
Exception {
  Object currentActivityThread =
  .invokeStaticMethod("android.app.ActivityThread",
  "currentActivityThread");

  // Get the original sPackageManager field of
  ActivityThread
  Object sPackageManager = RefInvoke.getFieldObject(
  currentActivityThread, "sPackageManager");

  // Prepare a proxy object to replace the original
  object
  Class<?> iPackageManagerInterface = Class.
  forName("android.content.pm.IPackageManager");
  Object proxy = Proxy.newProxyInstance(iPackageMana
  gerInterface.getClassLoader(),
```

```
    new Class<?>[] { iPackageManagerInterface },
    new MockClass3(sPackageManager));

  // Replace the sPackageManager field of
  ActivityThread
  RefInvoke.setFieldObject(currentActivityThread,
  "sPackageManager", proxy);
  }
}
```

The logic in *MockClass3* is shown as follows:

```
public class MockClass3 implements InvocationHandler {

  private Object mBase;

  public MockClass3(Object base) {
    mBase = base;
  }

  @Override
  public Object invoke(Object proxy, Method method,
  Object[] args) throws Throwable {
    if (method.getName().equals("getPackageInfo")) {
      return new PackageInfo();
    }
    return method.invoke(mBase, args);
  }
}
```

Place the method *hookPackageManage()*into *MockClass2* and invoke it in the method *handleLaunchActivity()* of *MockClass2*.

Let's recall the plug-in solution of *Activity* described earlier in this section. We put the object *LoadedApk* corresponding to the plug-in *apk* into the cache directly, and then change the *ClassLoader* of the object *LoadedApk* to the *ClassLoader* of the plug-in.

The disadvantage of this solution is obvious. We must use reflection to get a lot of instances of the internal class in the Android system, and we must we must write a lot of *if…else…* statements to resolve the code change among the different versions of the Android system.

8.2.3 Solution 2: Merge All the Plug-In *Dexes* into One Array*

Let's have a look at how to load the *dex* of the plug-in:

```
File extractFile = this.
getFileStreamPath(apkName);
String dexpath = extractFile.getPath();
File fileRelease = getDir("dex", 0); //0 means
Context.MODE_PRIVATE
DexClassLoader classLoader = new
DexClassLoader(dexpath,
    fileRelease.getAbsolutePath(), null,
    getClassLoader());
```

The variable *dexpath* is the path of the plug-in app. The Android system handles *dexPath* in *BaseDexClassLoader* and *DexPathList*:

```
public class BaseDexClassLoader extends ClassLoader {
  private final DexPathList pathList;

  public BaseDexClassLoader(String dexPath, File
  optimizedDirectory,
      String librarySearchPath, ClassLoader parent) {
  super(parent);

    this.pathList = new DexPathList(this, dexPath,
      librarySearchPath, optimizedDirectory);
  }
}

public class DexPathList {
  private Element[] dexElements;

  public DexPathList(ClassLoader definingContext,
  String dexPath,
      String libraryPath, File optimizedDirectory) {
    this.dexElements = makeDexElements(splitDexPath(de
    xPath), optimizedDirectory);
  }
```

* Sample code: https://github.com/BaoBaoJianqiang/ActivityHook2

```
private static List<File> splitDexPath(String path) {
  return splitPaths(path, false);
}

private static List<File> splitPaths(String
searchPath, boolean directoriesOnly) {
  List<File> result = new ArrayList<>();

  if (searchPath != null) {
for (String path : searchPath.split(File.
pathSeparator)) {
  //omit some codes....
    }
  }
 }
}
```

We split the string *dexPath* into an array by *File.pathSeparator* and each item in the string array is a path from the *dex/apk* in the HostApp or the plug-in, shown as follows:

```
/data/user/0/jianqiang.com.activityhook1/files/
plugin1.apk:/data/user/0/jianqiang.com.activityhook1/
files/plugin1.apk:
```

The string array will be converted into the field *dexElements* of *DexPathList*.

Based on this knowledge, we can manually add the *dex* of the plug-in into the array *dexElements* of the HostApp using hook technology.

There are three steps.

1) In the HostApp, get the field *dexElements* of *ClassLoader* of the HostApp:

  - Get the field *pathList* of *BaseDexClassLoader*; its type is *DexPathList*.

  - Get the field *dexElements* of *DexPathList*; it is an array.

2) Get an *Element* object from the variable *apkFile* of the plug-in.

3) Merge *dexElements* of the plug-in and the HostApp into a new *dex* array and replace the original field *dexElements* of the HostApp.

To carry out these three points, the code is as follows:

```
public static void patchClassLoader(ClassLoader cl,
File apkFile, File optDexFile)
    throws IllegalAccessException,
    NoSuchMethodException, IOException,
    InvocationTargetException,
    InstantiationException, NoSuchFieldException {
    // Get BaseDexClassLoader : pathList
    Object pathListObj = RefInvoke.
    getFieldObject(DexClassLoader.class.
    getSuperclass(), cl, "pathList");

    // Get PathList: Element[] dexElements
    Object[] dexElements = (Object[]) RefInvoke.
    getFieldObject(pathListObj, "dexElements");

    // Element type
    Class<?> elementClass = dexElements.getClass().
    getComponentType();

    // Create an array to replace the original array
    Object[] newElements = (Object[]) Array.
    newInstance(elementClass, dexElements.length + 1);

    // Plug-In constructor of Element(File file,
    boolean isDirectory, File zip, DexFile dexFile)
    Class[] p1 = {File.class, boolean.class, File.
    class, DexFile.class};
    Object[] v1 = {apkFile, false, apkFile, DexFile.
    loadDex(apkFile.getCanonicalPath(), optDexFile.
    getAbsolutePath(), 0)};
    Object o = RefInvoke.createObject(elementClass,
    p1, v1);

    Object[] toAddElementArray = new Object[] { o };
    // Copy the original elements
    System.arraycopy(dexElements, 0, newElements, 0,
    dexElements.length);
    // Copy the Plug-In elements
    System.arraycopy(toAddElementArray, 0,
    newElements, dexElements.length,
    toAddElementArray.length);

    // Replace
    RefInvoke.setFieldObject(pathListObj,
    "dexElements", newElements);
}
```

8.2.4 Plug-In Solution of *Resources**

Solutions 1 and 2 don't support resources in an *Activity*, but we have introduced how to load the resources of plug-ins in Section 8.1.4, and we can merge these solutions together.

The demo ZeusStudy1.2 is easy; we won't spend much time introducing it in detail; please read it for yourself.

8.2.5 Support *LaunchMode* in Plug-In†

Although we have introduced two solutions for the plug-in technique of *Activity*, we find all the *Activities* in the plug-ins have the same value as *LaunchMode* as *standard*, which is the default value of *LaunchMode*.

There are three other values of *LaunchMode*, *singleTop*, *singleTask*, and *singleInstance*.

If we have an *Activity* with *LaunchMode* = *singleTop* declared in the plug-in's *AndroidManifest.xml*, *singleTop* won't work at all, because the *AMS* will treat it as a *StubActivity* with the default value of *LaunchMode* as *standard*, not *singleTop*.

The best solution for these three *LaunchModes* is to supply a lot of *StubActivities*, some for *singleTop*, some for *singleTask*, and some for *singleInstance*, as shown in Figure 8.3.

FIGURE 8.3   Use stub to support different *LaunchMode*.

---

* Sample code: https://github.com/BaoBaoJianqiang/ZeusStudy1.2
† Sample code: https://github.com/BaoBaoJianqiang/ZeusStudy1.3

Next, we need to map between the real *Activities* of the plug-in and the *StubActivities* of the HostApp; for example, between *ActivityA* and *SingleTopActivity1*. We can define the mapping in a *JSON* file and download it from the remote server. In this book, we define mock data to simulate this scenario and store it in the collection *pluginActivies* of *MyApplication*. The code is as follows, *ActivityA* will have the same *LaunchMode* with *SingleTopActivity1*:

```
public class MyApplication extends Application {
  public static HashMap<String, String> pluginActivies
  = new HashMap<String, String>();

  void mockData() {
    pluginActivies.put("jianqiang.com.plugin1.
    ActivityA", "jianqiang.com.hostapp.
    SingleTopActivity1");
    pluginActivies.put("jianqiang.com.plugin1.
    TestActivity1", "jianqiang.com.hostapp.
    SingleTaskActivity2");
  }
}
```

Do you still remember *MockClass1*? This class intercepts the request of the method *startActivity()*, and we can add some logic to it to judge if the *LaunchMode* of the *Activity* of the plug-in is *default*, the code is as follows:

```
class MockClass1 implements InvocationHandler {

  @Override
  public Object invoke(Object proxy, Method method,
  Object[] args) throws Throwable {

    Log.e("bao", method.getName());

    if ("startActivity".equals(method.getName())) {

      //..omit some code

      String rawClass = raw.getComponent().
      getClassName();
```

```
if (MyApplication.pluginActivies.
containsKey(rawClass)) {
  String activity = MyApplication.pluginActivies.
  get(rawClass);
  int pos = activity.lastIndexOf(".");
  String pluginPackage = activity.substring(0,
  pos);
  componentName = new
  ComponentName(pluginPackage, activity);
} else {
  componentName = new ComponentName(stubPackage,
  StubActivity.class.getName());
}

//..omit some code
    }
  }
}
```

Let's test whether the mechanism of *LaunchMode* can work normally.

In the project Plugin1, the *LaunchMode* of *ActivityA* is *singleTop*, and the *LaunchMode* of *TestActivity1* is *singleTask*.

1) Click the button in *TestActivity1* to navigate to *ActivityA*, and then click the button "Goto TestActivity1" in *ActivityA* to navigate to *TestActivity1* again, Because the *LaunchMode* of *TestActivity1* is *singleTask*, the instance of *TestActivity1* created previously in the stack can be found, and the Android system won't create a new instance of *TestActivity1* but will go back to the previous instance of *TestActivity1*. This means the instance of *ActivityA* will be destroyed.

2) Click the button in *TestActivity1* to navigate to *ActivityA*. In *ActivityA*, click the button "Goto ActivityA" to navigate to *ActivityA* again, because the *LaunchMode* of *ActivityA* is *singleTop*, the instance of *ActivityA* will be reused without creating a new instance of *ActivityA*.

This solution has a small bug that whenever the *Activity* is *singleTop or singleTask*, the method *onCreate()* of this *Activity* won't be invoked, but the method *onNewIntent()* will be invoked. So, we need to intercept the method *onNewIntent()* in *MockClass2* to replace the *StubActivity* with the original *Activity* of the plug-in. The code is as follows:

```
class MockClass2 implements Handler.Callback {

  Handler mBase;

  public MockClass2(Handler base) {
    mBase = base;
  }

  @Override
  public boolean handleMessage(Message msg) {

    switch (msg.what) {
      // The value of "LAUNCH_ACTIVITY" of
      ActivityThread is 100
      // Use reflection is the best way, we use Hard
      coded here for simplicity
      case 100:
        handleLaunchActivity(msg);
        break;
      case 112:
        handleNewIntent(msg);
        break;
    }

    mBase.handleMessage(msg);
    return true;
  }

  private void handleNewIntent(Message msg) {
    Object obj = msg.obj;
    ArrayList intents = (ArrayList)RefInvoke.
    getFieldObject(obj, "intents");

    for(Object object : intents) {
      Intent raw = (Intent)object;
      Intent target = raw.getParcelableExtra(AMSHookHel
      per.EXTRA_TARGET_INTENT);
```

```
    if(target != null) {
      raw.setComponent(target.getComponent());

      if(target.getExtras() != null) {
        raw.putExtras(target.getExtras());
      }

      break;
    }
  }
}

//..omit some code
}
```

We won't spend time verifying whether *singleInstance* works in the plug-in. That's your homework.

8.2.6 Solution 3: Hook *ClassLoader**

We have introduced two solutions in Sections 8.2.2 and 8.2.3 for launching an *Activity* from the plug-in:

1) Create a corresponding *ClassLoader* for each plug-in.

2) Merge all the *dexes* into one array.

Now let's talk about the third solution, which replaces the *ClassLoader* of the HostApp with our own *ZeusClassLoader* directly.

ZeusClassLoader can play the role of the *ClassLoader* of the HostApp, which is done in the constructor of *ZeusClassLoader*, as follows:

```
ZeusClassLoader classLoader = new
ZeusClassLoader(mBaseContext.getPackageCodePath(),
mBaseContext.getClassLoader());

class ZeusClassLoader extends PathClassLoader {
  private List<DexClassLoader> mClassLoaderList = null;

  public ZeusClassLoader(String dexPath, ClassLoader
  parent) {
```

---

* Sample code: https://github.com/BaoBaoJianqiang/ZeusStudy1.4

```
    super(dexPath, parent);

    mClassLoaderList = new ArrayList<DexClassLoa
    der>();
  }
}
```

The variable *mClassLoaderList* of *ZeusClassLoader* stores all the *ClassLoaders* of the plug-in. First the method *loadClass(String class, boolean resolve)* of *ZeusClassLoader* tries to use the *ClassLoader* of the HostApp to load the class. If the *ClassLoader* of the HostApp can't load this class, it will traverse the collection *mClassLoaderList* until it finds a suitable *ClassLoader* that can load this class.

The code in *ZeusClassLoader* is as follows:

```
class ZeusClassLoader extends PathClassLoader {
  private List<DexClassLoader> mClassLoaderList = null;

  public ZeusClassLoader(String dexPath, ClassLoader
  parent) {
    super(dexPath, parent);

    mClassLoaderList = new ArrayList<DexClassLoa
    der>();
  }

  /**
   * Add a Plug-In to the current classLoader
   */
  protected void addPluginClassLoader(DexClassLoader
  dexClassLoader) {
    mClassLoaderList.add(dexClassLoader);
  }

  @Override
  protected Class<?> loadClass(String className,
  boolean resolve) throws ClassNotFoundException {
    Class<?> clazz = null;
    try {
      // First look for parent classLoader, here is the
      system to help us create the classLoader, the
      target corresponds to the HostApp
      clazz = getParent().loadClass(className);
```

214 ■ Android App-Hook and Plug-In Technology

```
    } catch (ClassNotFoundException ignored) {

    }

    if (clazz != null) {
      return clazz;
    }

    //find the ClassLoader in the mClassLoaderList
    if (mClassLoaderList != null) {
      for (DexClassLoader classLoader :
      mClassLoaderList) {
        if (classLoader == null) continue;
        try {
          // we only look for the plug-in's own apk and
          don't need to check the parent to avoid multiple
          useless queries and improve performance.
          clazz = classLoader.loadClass(className);
          if (clazz != null) {
            return clazz;
          }
        } catch (ClassNotFoundException ignored) {

        }
      }
    }
    throw new ClassNotFoundException(className + " in
    loader " + this);
  }
}
```

The code in *PluginManager* is as follows:

```
public class PluginManager {
  public static volatile ClassLoader mNowClassLoader =
  null; // System ClassLoader
  public static volatile ClassLoader mBaseClassLoader
  = null; // System ClassLoader

  public static void init(Application application) {
    mBaseClassLoader = mBaseContext.getClassLoader();
    mNowClassLoader = mBaseContext.getClassLoader();

    ZeusClassLoader classLoader = new
    ZeusClassLoader(mBaseContext.getPackageCodePath(),
    mBaseContext.getClassLoader());
```

```
File dexOutputDir = mBaseContext.getDir("dex",
Context.MODE_PRIVATE);
final String dexOutputPath = dexOutputDir.
getAbsolutePath();

for(PluginItem plugin: plugins) {
  DexClassLoader = new DexClassLoader(plugin.
  pluginPath,
      dexOutputPath, null, mBaseClassLoader);
  classLoader.addPluginClassLoader(dexClassLoader);
}

PluginUtil.setField(mPackageInfo, "mClassLoader",
classLoader);
Thread.currentThread().setContextClassLoader(classLo
ader);
mNowClassLoader = classLoader;
  }
}
```

But this solution is also not perfect. Originally, we started an *Activity* in the plug-in app as follows:

```
Intent intent = new Intent();
String activityName = PluginManager.plugins.get(0).
packageInfo.packageName + ".TestActivity1";
intent.setClass(this, Class.forName(activityName));
startActivity(intent);
```

When the method *Class.forName()* is invoked, the Android system will throw an exception that the HostApp can't find the *Activity* of the plug-in. Because *Class.forName()* uses *BootClassLoader* to load class, it is not hooking.

So, we couldn't use the method *Class.forName()* in the plug-in. We use the method *getClassLoader()*, shown as follows:

```
Intent intent = new Intent();
String activityName = PluginManager.plugins.get(0).
packageInfo.packageName + ".TestActivity1";
intent.setClass(this, getClassLoader().
loadClass(activityName));
startActivity(intent);
```

Activity is different from the other three components. It is widely used in apps. *Activity* interacts with the app user directly, and it has three important features, as follows:

- More lifecycle functions
- *LaunchMode*
- *Resources*

We need to support these three features of *Activity* in the plug-in.

8.3 THE PLUG-IN SOLUTION FOR SERVICE

Service is an Android component running in a background process. There are two forms of launching a *Service*: *startService()* and *bindService()*. We must support these two mechanisms with plug-in techniques.

8.3.1 The Relationship Between *Service* and *Activity*

In Section 2.7, we have introduced the family of *Context*. *Activity* and *Service* are both descendants of *Context*. There are so many similarities between *Activity* and *Service*; for example, both of them start with the help of *Context* and communicate with the *AMS*, and then the *AMS* notifies the app which component to launch, and forwards the message through *ActivityThread* and *H* (Figure 8.4).

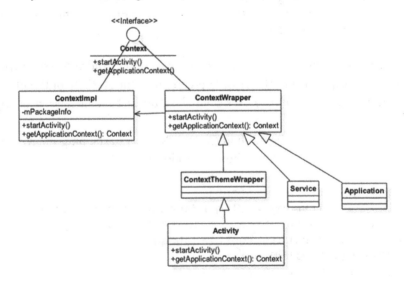

FIGURE 8.4   The family of *Context*.

There are also some differences between *Activity* and *Service*, let's have a look:

1) *Activity* is user-oriented; it has a lot of lifecycle methods; *Service* runs as a background process and has fewer lifecycle methods.

2) *Activity* has an important feature *LaunchMode*; there is a stack to store all the *Activities*. When we create a new *Activity*, it will be put on the top of the stack. The default value of *LaunchMode* is standard. If we create the instances of the same *Activity* many times, all the instances of this *Activity* will be put on the top of the stack. That's why we use *StubActivity* to cheat the *AMS* in plug-in programming. But *Service* doesn't support *LaunchMode*. When we invoke the method *startService()* of the same *Service* many times, only one instance of Service is generated. So, if we use *StubService* to cheat the *AMS*, it doesn't support multiple *Services* in the plug-in.

3) *ActivityThread* uses *Instrumentation* to launch an *Activity*. Instrumentation is not used for a *Service*.

There are two forms of starting a *Service*:

1) startService()

2) bindService()

The differences between these two forms are that the method *startService()* and the corresponding method *stopService()* are only used to start and stop a *Service*. For example, in a music app, we define a *Service*. We click the play button to notify the *Service* to execute the method *startService()* in the background processes to play the music.

When the method *bindService()* is executed, an instance of *ServiceConnection* will be sent to the *AMS*. The *AMS* will send a *Binder* object to the app; this object exists in the second parameter of the callback method *onServiceConnected()* of the *ServiceConnection*.

8.3.2 *StubService**

In the previous sections, we introduced that if we invoke the method *startService()* of the same *Service* several times, only one instance of the *Service* is created. So, we can't use only one *StubService* to hook multiple *Services* in the plug-in.

But we find the number of the *Services* in one app is less than ten; which is different from *Activity*. So, we can create ten *StubServices* in the HostApp, for example, *StubService1*, *StubService2*, and *StubService10*. Each *StubService* corresponds to only one *Service* of the plug-in, shown in Figure 8.5.

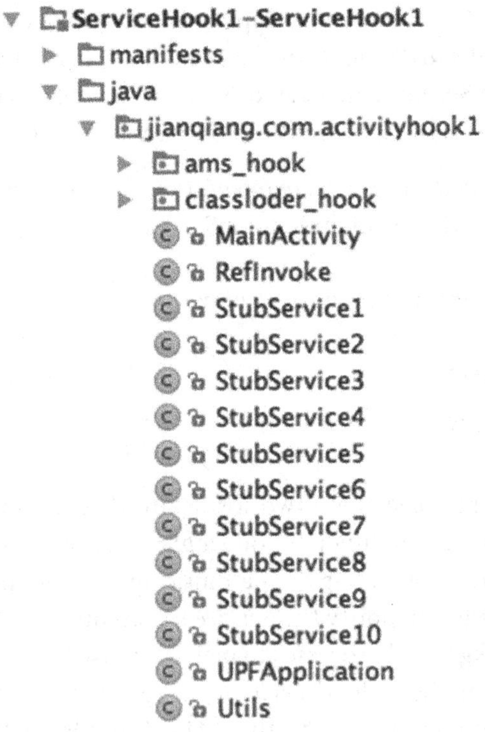

```
▼ 📂 ServiceHook1–ServiceHook1
   ▶ 📁 manifests
   ▼ 📁 java
      ▼ 📦 jianqiang.com.activityhook1
         ▶ 📦 ams_hook
         ▶ 📦 classloder_hook
            © MainActivity
            © RefInvoke
            © StubService1
            © StubService2
            © StubService3
            © StubService4
            © StubService5
            © StubService6
            © StubService7
            © StubService8
            © StubService9
            © StubService10
            © UPFApplication
            © Utils
```

FIGURE 8.5   Ten *StubServices* in the HostApp.

---

* Sample code: https://github.com/BaoBaoJianqiang/ServiceHook1

The next step is to set the mapping between *StubService* in the HostApp and the *Service* of the plug-in. There are two ways:

One way is to get a *JSON* string from the server; this file contains the 1:1 mapping, shown as follows.

```
{
  "plugins": [
    {
      "PluginService": "jianqiang.com.testservice1.
      MyService1",
      "StubService": "jianqiang.com.activityhook1.
      StubService1"
    },
    {
      "PluginService": "jianqiang.com.testservice1.
      MyService2",
      "StubService": "jianqiang.com.activityhook1.
      StubService2"
    }
  ]
}
```

Another way is to create a configuration file named *plugin_config* in the folder *Assets* of each plug-in app and put this *JSON* string into this file.

This second way is more natural.

The core code of parsing the configuration file *plugin_config* in the plug-in is as follows:

```
String strJSON = Utils.readZipFileString(dexFile.
getAbsolutePath(), "assets/plugin_config.json");
if(strJSON != null && !TextUtils.isEmpty(strJSON)) {
JSONObject jObject = new JSONObject(strJSON.
replaceAll("\r|\n", ""));
JSONArray jsonArray = jObject.getJSONArray("plugins");
for(int i = 0; i< jsonArray.length(); i++) {
  JSONObject jsonObject = (JSONObject)jsonArray.
  get(i);
  UPFApplication.pluginServices.put(
    jsonObject.optString("PluginService"),
    jsonObject.optString("StubService"));
  }
}
```

The method *readZipFileString* of *Utils* reads the contents of the configuration file *plugin_config* in the folder *Assets* of the plug-in*. It converts a *JSON* string into a *HashMap*. This *HashMap* is stored in the field *plugin-Services* of *UPFApplication* in the HostApp and it's a global variable. The field *pluginServices* plays a key role when a plug-in *Service* is loaded.

8.3.3 Plug-In Solution to *startService()*[†]

Plug-in solutions to *Service* and *Activity* are the same. Let's start with *startService()* and *stopService()*.

First, merge the *dexes* of the HostApp and the plug-in. We have introduced the *BaseDexClassLoaderHookHelper* in Section 8.2.3; we can use this class to load any class; don't worry about the *ClassNotFoundException* anymore.

Second, cheat the *AMS*. This implementation is in *AMSHookHelper*. The code is as follows:

```
public class AMSHookHelper {

  public static final String EXTRA_TARGET_INTENT =
  "extra_target_intent";

  public static void hookAMN() throws
  ClassNotFoundException,
    NoSuchMethodException, InvocationTargetException,
    IllegalAccessException, NoSuchFieldException {

    // Get the gDefault singleton of AMN, gDefault is
    final and static
    Object gDefault = RefInvoke.
    getStaticFieldObject("android.app.
    ActivityManagerNative", "gDefault");

    // gDefault is an instance of android.util.
    Singleton<T>. We get the mInstance field from this
    singleton
    Object mInstance = RefInvoke.
    getFieldObject("android.util.Singleton", gDefault,
    "mInstance");
```

---

* Refers from the ZeusPlugin framework: https://github.com/iReaderAndroid/ZeusPlugin
† Sample code: https://github.com/BaoBaoJianqiang/ServiceHook1

```
    // Create a proxy instance of MockClass1 and
    replace this field. Let the proxy object deal with
    it.
    Class<?> classB2Interface = Class.
    forName("android.app.IActivityManager");
    Object proxy = Proxy.newProxyInstance(
        Thread.currentThread().getContextClassLoader(),
        new Class<?>[] { classB2Interface },
        new MockClass1(mInstance));

    // Replace the mInstance field of mDefault with
    proxy object
    Class class1 = gDefault.getClass();
    RefInvoke.setFieldObject("android.util.Singleton",
    gDefault, "mInstance", proxy);
}

public static void hookActivityThread() throws
Exception {

    // Get the current ActivityThread object firstly
    Object currentActivityThread = RefInvoke.
    getStaticFieldObject("android.app.ActivityThread",
    "sCurrentActivityThread");

    // Since ActivityThread has only one process, we
    get the mH of this object
    Handler mH = (Handler) RefInvoke.getFieldObject(cu
    rrentActivityThread, "mH");

    // Replace mCallback field of Handler with new
    MockClass2(mH)
    RefInvoke.setFieldObject(Handler.class, mH,
    "mCallback", new MockClass2(mH));
}
}
```

Let's analyze *MockClass1* and *MockClass2* in the above code:

First, hook *AMN* and cheat the *AMS* to start *StubService*. The implementation of the code is in the class *MockClass1*. Now we try to intercept two methods *startService()* and *stopService()*. But we don't store the *Intent* in the cache. Because we can use *pluginServices* in *UPFApplication*, we can find the corresponding *StubService* according to the mapping defined in *pluginServices*:

```java
class MockClass1 implements InvocationHandler {

  private static final String TAG = "MockClass1";

  // replacing package name of StubService
  private static final String stubPackage =
  "jianqiang.com.activityhook1";

  Object mBase;

  public MockClass1(Object base) {
    mBase = base;
  }

  @Override
  public Object invoke(Object proxy, Method method,
  Object[] args) throws Throwable {

    Log.e("bao", method.getName());

    if ("startService".equals(method.getName())) {
      // Only intercept this method
      // Replace parameters as you want and even can
      replace the original Activity to start another
      Activity.
      // Find the first Intent object in the parameter
      int index = 0;
      for (int i = 0; i < args.length; i++) {
        if (args[i] instanceof Intent) {
          index = i;
          break;
        }
      }
    }

      //get StubService form UPFApplication.
      pluginServices
      Intent rawIntent = (Intent) args[index];
      String rawServiceName = rawIntent.getComponent().
      getClassName();
```

```
HashMap<String, String> a = UPFApplication.
pluginServices;

String stubServiceName = UPFApplication.
pluginServices.get(rawServiceName);

// replace Plug-In Service of StubService
ComponentName componentName = new
ComponentName(stubPackage, stubServiceName);
Intent newIntent = new Intent();
newIntent.setComponent(componentName);

// Replace Intent, cheat AMS
args[index] = newIntent;

Log.d(TAG, "hook success");
return method.invoke(mBase, args);
} else if ("stopService".equals(method.getName())) {
// Only intercept this method
// Replace parameters as you want and even can
replace the original Activity to start another
Activity.
// Find the first Intent object in the parameter
int index = 0;
for (int i = 0; i < args.length; i++) {
  if (args[i] instanceof Intent) {
    index = i;
    break;
  }
}

//Get StubService form UPFApplication.
pluginServices
Intent rawIntent = (Intent) args[index];
String rawServiceName = rawIntent.getComponent().
getClassName();
String stubServiceName = UPFApplication.
pluginServices.get(rawServiceName);

// Replace Plug-In Service of StubService
ComponentName componentName = new
ComponentName(stubPackage, stubServiceName);
```

```
   Intent newIntent = new Intent();
   newIntent.setComponent(componentName);

   // Replace Intent, cheat AMS
   args[index] = newIntent;

   Log.d(TAG, "hook success");
   return method.invoke(mBase, args);
 }

 return method.invoke(mBase, args);
 }
}
```

Second, after the *AMS* is cheated, it will inform the app to launch *StubService*. We need to hook the field *mCallback* of the *mH* and intercept the method *handleMessage()* of *ActivityThread*. We intercept the branch of 114(CREATE_SERVICE). This branch executes the method *handleCreateService()* of *ActivityThread*.

When the app starts a *Service*, the app sends an *Intent* object to *AMS*. We can't retrieve this Intent object of the method *handleCreateService()* of the class *ActivityThread*. The intent carries the information which service will be launched. We can get this information from the parameter data of the method *handleCreateService()*. The type of the parameter data is *CreateServiceData*, shown as follows:

```
private void handleCreateService(CreateServiceData
data) {
  LoadedApk packageInfo = getPackageInfoNoCheck(data.
  info.applicationInfo, data.compatInfo);
  Service service = null;

  java.lang.ClassLoader cl = packageInfo.
  getClassLoader();
  service = (Service) cl.loadClass(data.info.name).
   newInstance();

  // omit unrelated codes
  service.onCreate();
}
```

We replace *data.info.name* with *Service* in the plug-in, shown as follows:

```
class MockClass2 implements Handler.Callback {

  Handler mBase;

  public MockClass2(Handler base) {
    mBase = base;
  }

  @Override
  public boolean handleMessage(Message msg) {

    Log.d("baobao4321", String.valueOf(msg.what));
    switch (msg.what) {

      // The value of "CREATE_SERVICE" in
      ActivityThread is 114.
      // Use reflection is the best way, we use Hard
      coded here for simplicity
      case 114:
        handleCreateService(msg);
        break;
    }

    mBase.handleMessage(msg);
    return true;
  }

  private void handleCreateService(Message msg) {
    // For simplicity, take plug-in Service out
    directly

    Object obj = msg.obj;
    ServiceInfo serviceInfo = (ServiceInfo)RefInvoke.
    getFieldObject(obj, "info");

    String realServiceName = null;

    for (String key : UPFApplication.pluginServices.
    keySet()) {
      String value = UPFApplication.pluginServices.
      get(key);
```

```
   if(value.equals(serviceInfo.name)) {
     realServiceName = key;
     break;
   }
 }

 serviceInfo.name = realServiceName;
  }
}
```

Now a plug-in framework that supports the *startService()* is complete. Let's have a look how to use *MyService1* of the plug-in:

```
Intent intent = new Intent();
intent.setComponent(
  new ComponentName("jianqiang.com.testservice1",
  "jianqiang.com.testservice1.MyService1"));
startService(intent);
```

8.3.4 Plug-In Solution of *bindService**

Let's talk about another format of *Service*: *bindService()* and *unbindService()*.

With the help of the previous examples, the methods *bindService()* and *unbindService()* of *Service* are very simple. We can add some logic in *MockClass1* to cheat the *AMS* when the method *bindService()* is called.

```
else if ("bindService".equals(method.getName())) {

    // Find the first Intent object in the parameter
    int index = 0;
    for (int i = 0; i < args.length; i++) {
      if (args[i] instanceof Intent) {
        index = i;
        break;
      }
    }

    Intent rawIntent = (Intent) args[index];
    String rawServiceName = rawIntent.getComponent().
    getClassName();
```

---

* Sample code: https://github.com/BaoBaoJianqiang/ServiceHook2

```
    String stubServiceName = UPFApplication.
    pluginServices.get(rawServiceName);


    // Replace Plug-In Service of StubService
    ComponentName componentName = new
    ComponentName(stubPackage, stubServiceName);
    Intent newIntent = new Intent();
    newIntent.setComponent(componentName);

    // Replace Intent, cheat AMS
    args[index] = newIntent;

    Log.d(TAG, "hook success");
    return method.invoke(mBase, args);
  }
```

Next, we call *MyService2* of the plug-in in the HostApp:

```
findViewById(R.id.btnBind).setOnClickListener(new
View.OnClickListener() {
  @Override
  public void onClick(View v) {
    final Intent intent = new Intent();
    intent.setComponent(
      new ComponentName("jianqiang.com.testservice1",
        "jianqiang.com.testservice1.MyService2"));
    bindService(intent, conn, Service.
    BIND_AUTO_CREATE);
  }
});

findViewById(R.id.btnUnbind).setOnClickListener(new
View.OnClickListener() {
  @Override
  public void onClick(View v) {
    unbindService(conn);
  }
});
```

Up until now, we may have some questions.

- Why not cheat the *AMS* when *unbindService()* is invoked?

- Why not write code in *MockClass2* to replace *StubService2* with *MyService2*?

These two questions have also been bothering me for a long time.

The answer to the first question is simple. The method *bindService()* has a parameter *conn*. The type of *conn* is *ServiceConnection. conn* will be sent to *AMS* and stored in the *AMS* process. Now let's review the syntax of *unbindService()*. This method also has a parameter *conn*. The app process will send this parameter to *AMS*. If this *conn* was stored in the *AMS* process before, *AMS* will find it. That's why we needn't cheat *AMS* when the app executes the method *unbindService()* of the service in the plug-in app.

The second question is also very interesting. The flowchart of *bindService()* is shown in Figure 8.6 (the *AMS* sends a message to the app processes).

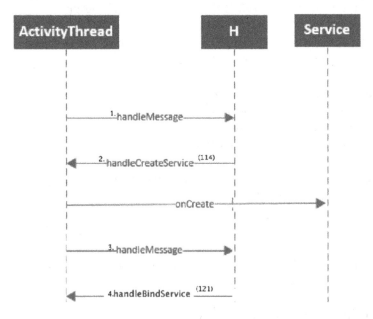

FIGURE 8.6   Flowchart of *bindService()*.

In other words, the method *bindService()* goes through branch 114 (the method *handleCreateService()*) first and then goes through branch 121 (the method *handleBindService()*).

The method *handleCreateService()* puts *MyService2* into the collection *mServices*.

Then in the methods *handleBindService()* (branch 121) and *handleUnbindService()* (branch 122) we can find both *MyService2* in the collection *mService* and execute the corresponding methods *bindService()* and *unbindService()*.

We intercept branch 114 to resolve the logic of *createService()*. In the method *handleCreateService()*, we switch *ServiceName* from *StubService2* to *MyService2*. Therefore, we do not need to intercept branch 121 and 122 and add and logic in *MockClass2*.

In this section, we have introduced a plug-in solution for *Service*. This solution requires that we pre-declare ten *StubServices* in the *AndroidManifest.xml* of the HostApp. When the number of services in the plug-in is more than ten, this solution is not suitable.

In Chapter 9, we will introduce another plug-in framework, *DL*, written by Yugang Ren. *DL* will give a final plug-in solution for *Service*.

8.4 A PLUG-IN SOLUTION FOR BROADCASTRECEIVER

BroadcastReceiver (*Receiver* for short in this book) is the simplest component in Android. It is a class that implements the design pattern Observer.

We have introduced the principle of *Receiver* in detail in Chapter 2. We talk about the plug-in solutions for the *Receiver* in this section. The *Receiver* is divided into two kinds: *Dynamic Receiver* and *Static Receiver*, the implementations of these two kinds are different.

8.4.1 *Receiver* Overview

The *Receiver* is divided into two kinds: *Static Receiver* and *Dynamic Receiver*. Let's have a look at the differences between them briefly:

1) *Static Receiver* must be declared in the *AndroidManifest.xml*. The *PMS* will parse the *AndroidManifest.xml* of the app after the user downloads and installs an app; this means all the *Static Receivers* exist in *PMS*.

2) *Dynamic Receiver* is registered by the app developer in the code; the method *registerReceiver()* of *Context* calls the method *AMN.getDefault().RegisterReceiver*, so all the *Dynamic Receivers* exist in the *AMS*.

The only difference between *Static Receiver* and *Dynamic Receiver* is the registration mechanism discussed above. *Static Receiver* and *Dynamic Receiver* have the same format for sending and receiving broadcasts.

1) Send a broadcast to the *AMS* using the method *sendBroadcast()* of *Context*, and it will invoke the method *AMN.getDefault(). broadcastIntent.*

2) After receiving the above information, the *AMS* will search the broadcasts stored in the *AMS* and the *PMS* for which ones meet the conditions, and then notify the App process to start these receivers and invoke the method *onReceive()* of these receivers.

The lifecycle of the *Receiver* is very simple. There is only one lifecycle method *onReceive()*, which is a callback function.

When *Receiver* sends or receives a broadcast, it will carry an *intent-filter*. For example, we need set an *intent-filter* when we want to send broadcasts, and then the *AMS* will know which broadcasts are suitable for the *intent-filter*. The code as follows:

1) Static Receiver declared in the AndroidManifest.xml:

```
<receiver
  android:name=".MyReceiver"
  android:enabled="true"
  android:exported="true">
  <intent-filter>
    <action android:name="baobao2" />
  </intent-filter>
</receiver>
```

2) Register *Dynamic Receiver* in the code:

```
MyReceiver myReceiver = new MyReceiver();
IntentFilter intentFilter = new IntentFilter();
intentFilter.addAction("baobao2");
registerReceiver(myReceiver, intentFilter);
```

3) Send a broadcast:

```
Intent intent = new Intent();
intent.setAction("baobao2");
intent.putExtra("msg", "Jianqiang");
sendBroadcast(intent);
```

8.4.2 A Plug-In Solution for *Dynamic Receiver**

Dynamic Receiver needn't communicate with the *AMS*, so it is only a normal class.

We only make sure that the HostApp can load this normal class in the plug-in. In Section 8.2.2, we introduced the technique of *dex* merging. We can use *BaseDexClassLoaderHookHelper* in the project Receiver1.0, and then the *Dynamic Receiver* in the plug-in can be invoked normally.

8.4.3 A Plug-In Solution for *Static Receiver*†

Static Receiver must be declared in the *AndroidManifest.xml*, which is similar to *Activity*.

In Section 8.2, we introduced a plug-in solution for *Activity*. *Activity* is not declared in the *AndroidManifest.xml* and can be launched. We use *StubActivity* to cheat the *AMS*.

But this is not suitable for *Static Receiver*, because we must specify an *intent-filter* for the broadcast. The parameters of the *intent-filter* such as *action* can be set to any value, so we cannot use a *Stub* mechanism in the *Receiver* directly.

However, we have another solution. We try to treat *Static Receiver* declared in the *AndroidManifest.xml* of the plug-in as a *Dynamic Receiver*, as follows:

Step 1: Because the *PMS* can only read the *AndroidManifest.xml* of the HostApp, parse, and register *Static Receiver*. We can use the *PMS* to parse the *Static Receiver* declared in the *AndroidManifest.xml* of the plug-in by reflection.

Step 1: Traverse the list of *Static Receivers* we get in step 1. Use the *ClassLoader* in the plug-in to load each *Receiver* in the list, instantiate them as an object, and register them in the *AMS* as *Dynamic Receivers*.

The *PMS* parses the *AndroidManifest.xml* using the method *PackageParser()*. The definition of the method *PackageParser()* is as follows:

```
public Package parsePackage(File packageFile, int
flags)
```

* Sample code: https://github.com/BaoBaoJianqiang/Receiver1.0
† Sample code: https://github.com/BaoBaoJianqiang/Receiver1.1

Let's analyze the parameters and return a value for this method:

- The first parameter is *apk*, which can be specified as a plug-in *apk*;

- The second parameter is a filter. When we set it to *PackageManager. GET_RECEIVERS*, all the *Static Receivers* in the *apk* will be returned;

- The return value is a *Package* object, storing the *Static Receivers* we get from the *AndroidManifest.xml*.

So, we can get a *Package* object by using the method *parsePackage()*, this *Package* object represents for a plug-in app. We can get all the *Static Receivers* of this plug-in app from this *Package* object. The code is as follows:

```
// First call parsePackage to get the
corresponding Package object of the Apk object
Object packageParser = RefInvoke.
createObject("android.content.pm.PackageParser");
Class[] p1 = {File.class, int.class};
Object[] v1 = {apkFile, PackageManager.
GET_RECEIVERS};
Object packageObj = RefInvoke.invokeInstanceMethod
(packageParser, "parsePackage", p1, v1);

// Read the receivers field in the Package. It is
a List<Activity>
// The next thing to do is to get the ActivityInfo
corresponding to the Receiver according to this
List<Activity> (still regard the receiver
information as the activity information)
List receivers = (List) RefInvoke.
getFieldObject(packageObj, "receivers");
for (Object receiver : receivers) {
  registerDynamicReceiver(context, receiver);
}
```

In the *for* loop of the above code, we invoke the method *registerDynalmicReceiver()* to convert each *Static Receiver* to a *Dynamic Receiver* and register all these *Receivers* in the *AMS*. The implementation is as follows

```
// Parse the receiver and the corresponding
intentFilter
// Register Receiver manually
public static void registerDynamicReceiver(Context
context, Object receiver) {
  // Get the intents field of receiver
  List<? extends IntentFilter> filters = (List<?
  extends IntentFilter>) RefInvoke.getFieldObject(
      "android.content.pm.PackageParser$Component",
      receiver, "intents");

  try {
    // Register each static Receiver as dynamic
    for (IntentFilter intentFilter : filters) {
      ActivityInfo receiverInfo = (ActivityInfo)
      RefInvoke.getFieldObject(receiver, "info");

      BroadcastReceiver broadcastReceiver =
      (BroadcastReceiver) RefInvoke.
      createObject(receiverInfo.name);
      context.registerReceiver(broadcastReceiver,
      intentFilter);
    }
  } catch (Exception e) {
    e.printStackTrace();
  }
}
```

8.4.4 A Final Plug-In Solution for *Static Receiver**

Static Receiver has a feature enabling it to launch even if the app is not launched because the *PMS* has read all the *Static Receivers* in the *AndroidManifest.xml* before the app is launched.

The solution introduced in Section 8.4.3 is not the best solution. In this solution, we convert all the *Static Receivers* from the plug-in into *Dynamic Receivers*. This means we must launch the app first; otherwise, the *Static Receivers* from the plug-in won't be launched.

Most app developers misunderstand that "launching an app is to launch an *Activity*." *Activity* is one of the four Android components. We

---

* Sample code: https://github.com/BaoBaoJianqiang/Receiver1.2

can launch a *Service* or send a broadcast to the *Receiver* before the home page of the app is launched.

So, we continue to explore how to send a broadcast to the *Static Receiver* from the plug-in without launching the app.

We introduced *StubActivity* and *StubService* in Section 8.2 and 8.3; now let's review this technique:

- All the *Activities* can be mapped to only one *StubActivity*. If the *LaunchMode* is not a *standard*(default value), we need more *StubActivities*.

- Each *Service* in the plug-in app corresponds to only one *StubService*. But a *StubService* in the HostApp can be used for a *Service* from the plug-in. We will introduce a 1:n plug-in solution in Section 9.2.4.

We can also create a lot of *StubReceivers* and create 1:1 mapping between the *StubReceiver* and *Receiver* from the plug-in.

1) *Receiver* has an interesting feature: *Action*. Each *Receiver* carries one or more *Actions*. So, we can create only one *StubReceiver* but add many *Actions* for this *StubReceiver*, and each *Action* is mapped to the *Receiver* of the plug-in, as follows:

```
<receiver
  android:name=".StubReceiver"
  android:enabled="true"
  android:exported="true">
    <intent-filter>
      <action android:name="jianqiang1" />
    </intent-filter>
    <intent-filter>
      <action android:name="jianqiang2" />
    </intent-filter>
    <intent-filter>
      <action android:name="jianqiang3" />
    </intent-filter>
    <intent-filter>
      <action android:name="jianqiang4" />
    </intent-filter>
</receiver>
```

In this *StubReceiver*, there are four *Actions*, the name of each *Action* is "jianqiang1," "jianqiang2," "jianqiang3," and "jianqiang4."

2) In the *AndroidManifest.xml*, there is a *meta-data* tag for each component. We can configure this tag for the *Services* from the plug-in; for example, *MyReceiver* has a *meta-data* tag, its name is "oldAction" and its value is "jianqiang," shown as follows:

```xml
<?xml version="1.0" encoding="utf-8"?>
<manifest xmlns:android="http://schemas.android.com/
apk/res/android"
  package="jianqiang.com.receivertest">

  <application
    android:allowBackup="true"
    android:icon="@mipmap/ic_launcher"
    android:label="@string/app_name"
    android:supportsRtl="true"
    android:theme="@style/AppTheme">

    <receiver
      android:name=".MyReceiver"
      android:enabled="true"
      android:exported="true">
      <intent-filter>
        <action android:name="baobao" />
      </intent-filter>
      <meta-data android:name="oldAction" android:val
      ue="jianqiang1"></meta-data>
    </receiver>
    <receiver
      android:name=".MyReceiver2"
      android:enabled="true"
      android:exported="true">
      <intent-filter>
        <action android:name="baobao2" />
      </intent-filter>
      <meta-data android:name="oldAction" android:val
      ue="jianqiang2"></meta-data>
    </receiver>
  </application>
</manifest>
```

This plug-in has two *Static Receivers*, *MyReceiver* has the *Action* "bao-bao" and meta-data tag "oldAction=jianqiang," and *MyReceiver2* has the *Action* "baobao2" and meta-data tag "oldAction=jianqiang2"; we convert these two *Static Receivers* to *Dynamic Receivers*, and store the mapping "jianqiang:baobao" and "jianqiang2:baobao2" in a *HashMap*. The code is as follows (we use *ReceiverManager.pluginReceiverMappings* to store all the mapping):

```
public final class ReceiverHelper {
  private static final String TAG =
  "ReceiverHelper";

  /**
   * Parse the <receiver> in the Plug-In Apk file
   and store it
   *
   * @param apkFile
   * @throws Exception
   */
  public static void preLoadReceiver(Context
  context, File apkFile) {
    // First, call parsePackage to get the Package
    object of the Apk
    Object packageParser = RefInvoke.
    createObject("android.content.pm.PackageParser");
    Class[] p1 = {File.class, int.class};
    Object[] v1 = {apkFile, PackageManager.
    GET_RECEIVERS};
    Object packageObj = RefInvoke.invokeInstanceMetho
    d(packageParser, "parsePackage", p1, v1);

    String packageName = (String)RefInvoke.
    getFieldObject(packageObj, "packageName");

    // Read the receivers field in the Package. This
    is a List<Activity>
      // The next thing to do is to get the
      ActivityInfo corresponding to the Receiver
      according to this List<Activity> (still regard
      the receiver information as the activity
      information)
```

```
List receivers = (List) RefInvoke.
getFieldObject(packageObj, "receivers");

try {
  for (Object receiver : receivers) {
    Bundle metadata = (Bundle)RefInvoke.
    getFieldObject(
        "android.content.
        pm.PackageParser$Component", receiver,
        "metaData");
    String oldAction = metadata.
    getString("oldAction");

    // Parse the receiver and the corresponding
    intentFilter
    List<? extends IntentFilter> filters =
    (List<? extends IntentFilter>) RefInvoke.
    getFieldObject("android.content.
    pm.PackageParser$Component", receiver,
    "intents");

    // Register each static Receiver as dynamic
    for (IntentFilter intentFilter : filters) {
      ActivityInfo receiverInfo = (ActivityInfo)
      RefInvoke.getFieldObject(receiver, "info");
      BroadcastReceiver broadcastReceiver =
      (BroadcastReceiver) RefInvoke.
      createObject(receiverInfo.name);
      context.registerReceiver(broadcastReceiver,
      intentFilter);

      String newAction = intentFilter.
      getAction(0);
      ReceiverManager.pluginReceiverMappings.
      put(oldAction, newAction);
    }
  }
} catch (Exception e) {
  e.printStackTrace();
  }
 }
}
```

We use the method *preLoadReceiver()* of *ReceiverHelper* to parse *AndroidManifest.xml*, fetch all the *Static Receivers* and get the tag *metadata* for each *Receiver*, we have introduced this method in Section 8.4.3.

3) According to this configuration, when we send a broadcast with the *intent-filter* "action=jianqiang1," the method *onReceive()* of *StubReceiver* will be fired; it searches the mapping collection *ReceiverManager.pluginReceiverMappings*, and finds the mapping "jianqiang:baobao," and then send a new broadcast with the *intent-filter* "action=baobao," finally the method *onReceive()* of the *Receiver* from the plug-in will be fired. The code of the *StubReceiver* is as follows:

```
public class StubReceiver extends BroadcastReceiver {
  public StubReceiver() {
  }

  @Override
  public void onReceive(Context context, Intent
  intent) {
    String newAction = intent.getAction();
    if(ReceiverManager.pluginReceiverMappings.
    containsKey(newAction)) {
      String oldAction = ReceiverManager.
      pluginReceiverMappings.get(newAction);
      context.sendBroadcast(new Intent(oldAction));
    }
  }
}
```

Up to now, we have introduced the plug-in solution for the *Static Receiver*. We can send a broadcast to the *Static Receiver* in the plug-in even if the app hasn't been launched.

This section introduces a plug-in solution for *Dynamic Receiver* and *Static Receiver*.

In Section 9.2.5, we will introduce a new plug-in solution for a *Receiver* based on *Static-Proxy*. This solution takes a *Receiver* from the plug-in as a normal class and invokes the method *onReceive()* of *ProxyReceiver* in the plug-in.

8.5 A PLUG-IN SOLUTION FOR CONTENTPROVIDER

This section introduces a plug-in solution for *ContentProvider*.

Some app developers may be not familiar with *ContentProvider* because it's not widely used in apps. Actually, it's only used in special scenarios. For example, when an app reads the mobile address book.

8.5.1 The Basic Concept of *ContentProvider*

ContentProvider is used to provide a large amount of data.

ContentProvider is not widely used in apps.

ContentProvider is widely used in the custom Android ROM such as MIUI. The communication between different components always uses *ContentProvider* to transport large amounts of data.

ContentProvider is an *SQLite* database, which is divided into data provider and data user. They transmit data via *Anonymous Shared Memory (ASM)*. For example, there is a data provider named A, and a data user named B. B asks A for data and tells A "you can write the data in this memory address"; A prepares the data, writes in the memory address required by B, and then B can use this data directly. *ASM* is different from traditional data transmission; it's suitable for transmitting large amounts of data.

However, *ContentProvider* is not suitable in all scenarios, for example, when the app jumps from *ActivityA* to *ActivityB* and transfers some data, and the data is too small, such as a string or an integer. The data transmission is based on *Binder*, and *Binder* communicates quickly with the process.

It's necessary to use *ContentProvider* when the data is more than 1M; otherwise, *Binder* is enough.

8.5.2 A Simple Example of *ContentProvider**

Let's write an example to help us understand *ContentProvider*. There are two apps, A1 and B1:

There is a *ContentProvider* defined in B1 to provide data, and A1 calls the *ContentProvider* of B1 to obtain data.

The *ContentProvider* declared in B1:

```
<provider
    android:name=".MyContentProvider"
```

---

* Sample code: https://github.com/BaoBaoJianqiang/ContentProvider1

```
android:authorities="baobao"
android:enabled="true"
android:exported="true"/>
```

Now A1 can visit the *ContentProvider* of B1 by using the following URI:

```
content://baobao/
```

Let's have a look at *MyContentProvider* defined in B1:

```
public class MyContentProvider extends ContentProvider {
  public MyContentProvider() {
  }

  @Override
  public boolean onCreate()
  {
    System.out.println("===onCreate===");
    return true;
  }

  @Override
  public int delete(Uri uri, String where, String[]
  whereArgs)
  {
    System.out.println(uri + "===delete===");
    System.out.println("where:" + where);
    return 1;
  }

  @Override
  public Cursor query(Uri uri, String[] projection,
  String where,
          String[] whereArgs, String sortOrder)
  {
    // Omit a lot of code
  }

  @Override
  public Uri insert(Uri uri, ContentValues values)
  {
    // Omit a lot of code
  }
```

```
@Override
public int update(Uri uri, ContentValues values,
String where,
        String[] whereArgs)
{
  // Omit a lot of code
}
}
```

All the *ContentProviders* must implement CRUD methods. For example, in the *ContentProvider* of B1, I implemented the method *delete()*. The implementation is simple, it returns directly.

Now, let's have a look at how to use the *ContentProvider* of B1 in A1.

```
public class MainActivity extends Activity {

ContentResolver contentResolver;
Uri uri;

@Override
protected void onCreate(Bundle savedInstanceState) {
  super.onCreate(savedInstanceState);
  setContentView(R.layout.activity_main);

  uri = Uri.parse("content://baobao/");
  contentResolver = getContentResolver();
}

public void delete(View source) {
  int count = contentResolver.delete(uri, "delete_
  where", null);
  Toast.makeText(this, "delete uri:" + count, Toast.
  LENGTH_LONG).show();
}
}
```

In A1, we invoke the method *getContentResolver()* to get the handle of *ContentProvider* and invoke its method *delete()* to invoke the method *delete()* in the remote *ContentProvider* of B1. We need to specify the URI of the remote *ContentProvider* in the CRUD method; in this demo the URI is "content://baobao/."

Install A1 and B1 on the Android phone; it doesn't matter if you launch B1 or not.

When we click the delete button in A1, the method delete() of MyContentProvider is invoked and this method returns 1.

8.5.3 A Plug-In Solution for *ContentProvider**

Do you still remember the plug-in solution for *BroadcastReceiver*? All the *Static Receivers* of the plug-in are converted to *Dynamic Receivers*, and then they are registered manually in the *Receiver* collection of the HostApp.

Actually, *ContentProvider* has the same plug-in implementation as *BroadcastReceiver*. But it is called "installation," rather than "registration."

The logic of "installation" is in the method *installContentProviders()* of *ActivityThread*, as follows:

```
private void installContentProviders(
  Context context, List<ProviderInfo> providers) {
final ArrayList<IActivityManager.
ContentProviderHolder> results =
  new ArrayList<IActivityManager.
  ContentProviderHolder>();

for (ProviderInfo cpi : providers) {
  IActivityManager.ContentProviderHolder cph =
  installProvider(context, null, cpi,
    false /*noisy*/, true /*noReleaseNeeded*/,
    true /*stable*/);
  if (cph != null) {
    cph.noReleaseNeeded = true;
    results.add(cph);
  }
}

try {
  ActivityManagerNative.getDefault().
  publishContentProviders(
    getApplicationThread(), results);
} catch (RemoteException ex) {
}
}
```

* Sample code: https://github.com/BaoBaoJianqiang/ContentProvider2

We have the chance to execute this method manually and put the *ContentProvider* of the plug-in as the second parameter of this method.

Now we get the plug-in solution for *ContentProvider*, shown as follows:

1) Merge the *dex* of the HostApp and the plug-in. Refer to the *BaseDexClassLoaderHookHelper* class. We introduced this technique in Section 8.1.2.

2) Invoke the method *parsePackage()* of *PackageParser* to fetch the information from the *ContentProvider* of the plug-in; this method returns a *Package* object, and we use the method *generateProviderInfo()* to convert it to the *ProviderInfo* object we need.

```
public static List<ProviderInfo> parseProviders(File
apkFile) throws Exception {

  // get an instance of the PackageParser object
  Class<?> packageParserClass = Class.
  forName("android.content.pm.PackageParser");
  Object packageParser = packageParserClass.
  newInstance();

  // First,execute parsePackage method to get
  corresponding Package object for the Apk object.
  Class[] p1 = {File.class, int.class};
  Object[] v1 = {apkFile, PackageManager.
  GET_PROVIDERS};
  Object packageObj = RefInvoke.invokeInstanceMethod
  (packageParser, "parsePackage",p1, v1);

  // Read the services field in the Package object
  // Next, get the corresponding ProviderInfo for
  Provider according to the List<Provider>.
  List providers = (List) RefInvoke.
  getFieldOjbect(packageObj.getClass(), packageObj,
  "providers");

  // execute the generateProviderInfo method to
  convert PackageParser.Provider to ProviderInfo

  // prepare the parameters that the
  generateProviderInfo method required
```

```
Class<?> packageParser$ProviderClass = Class.
forName("android.content.
pm.PackageParser$Provider");
Class<?> packageUserStateClass = Class.
forName("android.content.pm.PackageUserState");
Object defaultUserState = packageUserStateClass.
newInstance();
int userId = (Integer) RefInvoke.
invokeStaticMethod("android.os.UserHandle",
"getCallingUserId", null, null);
Class[] p2 = {packageParser$ProviderClass, int.
class, packageUserStateClass, int.class};

List<ProviderInfo> ret = new ArrayList<>();
// parse the Provider component according to the
Intent
for (Object provider : providers) {
  Object[] v2 = {provider, 0, defaultUserState,
  userId};
  ProviderInfo info = (ProviderInfo) RefInvoke.invo
  keInstanceMethod(packageParser,
  "generateProviderInfo",p2, v2);
  ret.add(info);
}

  return ret;
}
```

3) Put the *ContentProvider* of the plug-in in the *ContentProvider* collection of the HostApp, so that the HostApp can treat them as "real" *ContentProviders*, rather than a normal class. We also need set the file *packageName* of these *ContentProviders* to the *packageName* of the current *apk*. The code is as follows:

```
for (ProviderInfo providerInfo : providerInfos) {
providerInfo.applicationInfo.packageName = context.
getPackageName();
}
```

4) Invoke the method *installContentProviders()* of *ActivityThread* by reflection, and pass the *ContentProviders* of the plug-in as a parameter to this method:

```
Object currentActivityThread = RefInvoke.
invokeStaticMethod("android.app.ActivityThread",
"currentActivityThread", null, null);

Class[] p1 = {Context.class, List.class};
Object[] v1 = {context, providerInfos};

RefInvoke.invokeInstanceMethod(currentActivityThr
ead, "installContentProviders", p1, v1);
```

8.5.4 The Right Time to Install a *ContentProvider* Plug-In*

We have introduced a plug-in solution for *ContentProvider*. Sometimes when we invoke the CRUD method on the *ContentProvider* of the plug-in, we find this *ContentProvider* is still not "installed" in the HostApp.

So, we want to install the *ContentProvider* of the plug-in as soon as possible.

The method *installContentProviders()* of *ActivityThread* is used to install *ContentProvider*, which is invoked after the app process is launched; it's invoked earlier than the method *onCreate()* of *Application*, but later than the method *attachBaseContext()* of *Application*. So, we can execute the method *installContentProviders()* manually in the method *attachBaseContext()*, as follows:

```
public class UPFApplication extends Application {

@Override
protected void attachBaseContext(Context base) {
  super.attachBaseContext(base);

  try {
    File apkFile = getFileStreamPath("plugin2.apk");
    if (!apkFile.exists()) {
      Utils.extractAssets(base, "plugin2.apk");
    }

    File odexFile = getFileStreamPath("plugin2.
    odex");
```

---

* Sample code: https://github.com/BaoBaoJianqiang/ContentProvider2

```
    //Hook ClassLoader,ensure that the class in the
    Plug-In can be loaded successfully
    BaseDexClassLoaderHookHelper.patchClassLoader(get
    ClassLoader(), apkFile, odexFile);

    //install Plug-In Providers
    ProviderHelper.installProviders(base,
    getFileStreamPath("plugin2.apk"));
  }catch (Exception e) {
    throw new RuntimeException("hook failed", e);
  }
 }
}
```

8.5.5 The Forwarding Mechanism of *ContentProvider**

It is not an ideal solution that the third app invokes the *ContentProvider* of plug-in directly.

It is better to define a *StubContentProvider* in the HostApp. We export a *StubContentProvider* into the third app. When the third app invokes this *StubContentProvider*, we can forward *StubContentProvider* to the *ContentProvider* of the plug-in, as shown in Figure 8.7.

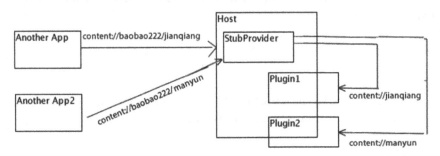

FIGURE 8.7 Forwarding *ContentProvider*.

Let's write a method *getRealUri()* to convert the URI, shown as follows:

```
private Uri getRealUri(Uri raw) {
    String rawAuth = raw.getAuthority();
    if (!AUTHORITY.equals(rawAuth)) {
      Log.w(TAG, "rawAuth:" + rawAuth);
    }
```

---

* Sample code: https://github.com/BaoBaoJianqiang/ContentProvider2

```
String uriString = raw.toString();
uriString = uriString.replaceAll(rawAuth + '/',
"");
Uri newUri = Uri.parse(uriString);
Log.i(TAG, "realUri:" + newUri);
return newUri;
}
```

The method *getRealUri()* will convert the URI protocol such as "content://host_auth/plugin_auth/path/query" to "content://plugin_auth/path/query."

For example, "content://baobao222/jianqiang" can be converted to "content://jianqiang."

ContentProvider is a database engine. It provides CRUD methods to the third app.

The plug-in solution *ContentProvider* is simple: read all the *ContentProviders* in the *AndroidManifest.xml* of the plug-in and put them into the *ContentProviders* collection of the HostApp.

The key point of this solution is the forwarding mechanism. We expose a *StubContentProvider* to the third app. The third app doesn't know the *ContentProviders* of the plug-in. What we need to do is forward the *StubContentProviders* to the *ContentProviders* of the plug-in using a different URI.

8.6 SUMMARY

This chapter introduced the core techniques for using plug-ins using the four components that make up the basic app.

We use a lot of hook techniques in this chapter; we modify the original behavior of some internal methods not open to app developers. In Chapter 9, we will introduce another solution with a little modification to these internal methods.

A Plug-In Solution
Based on Static-Proxy

9.1 A PLUG-IN SOLUTION FOR *ACTIVITY* BASED ON *STATIC-PROXY*

This chapter introduces my favorite plug-in framework designed by Yugang Ren*, *dynamic-load-apk* (*DL* for short). *DL* invents a new keyword "that," so this framework has an interesting alias "That."

In this chapter, I will introduce the design idea of the "That" framework in detail and write a new "That" framework from scratch, including the plug-in implementation of *Activity*, *Service*, and *BroadcastReceiver*.

9.1.1 The Idea of *Static-Proxy*

As we introduced in the previous chapters, when we create an instance of *Activity* in the plug-in using reflection, this instance doesn't have lifecycle methods; it's only a normal class object.

For example, we cheat the *AMS* not to check the *Activity* without declaring in the *AndroidManifest.xml*. But if we don't put the *Activity* from the Plug-In into the *Activity* collection of the HostApp, the method *onCreate()* of this *Activity* can't be invoked by the *AMS*, and the other lifecycle methods of *Activity* such as *onResume()*, *onPause()* have the same problem. Because the HostApp doesn't treat this *Activity* from the plug-in as a real *Activity*, it's only a normal class.

---

* https://github.com/singwhatiwanna/dynamic-load-apk

To invoke these methods normally, we create a proxy class *Proxy Activity* in the HostApp and declare it in the *AndroidManifest.xml*, so it's a real *Activity*.

ProxyActivity has a reference to *ActivityA* in the plug-in. In the logic of the method *onCreate()* of *ProxyActivity*, the method *onCreate()* of *ActivityA* is invoked, as shown in Figure 9.1.

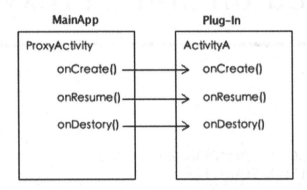

FIGURE 9.1   Core of the "That" framework.

This is similar to a puppet show. The artists pull the wire, and the puppets at the end of the wire move according to the pulling action. *ProxyActivity* is the artist, and *ActivityA* of the plug-in is the puppet. The puppet has no life. *ActivityA* of the plug-in is a normal class; it also doesn't have lifecycle methods.

9.1.2 The Simplest Example of *Static-Proxy**

In this section, I will introduce an example of the implementation of *Static-Proxy*. Figure 9.2 shows the most important classes in this example.

---

* Sample code: https://github.com/BaoBaoJianqiang/That1.0

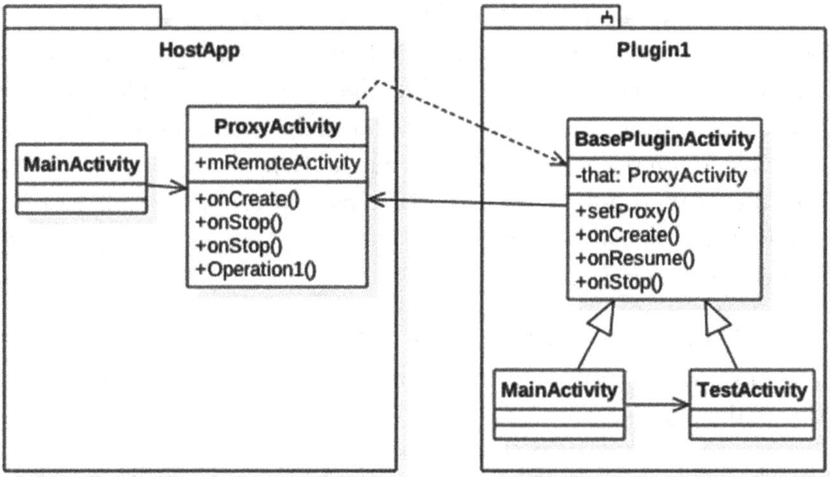

FIGURE 9.2 Relationship between classes of the HostApp and plug-in.

9.1.2.1 Jump from the HostApp to the Plug-In

In the HostApp, *ProxyActivity* is a proxy for all the *Activities* of the plug-in.
We can use *ProxyActivity* with different parameters to open any *Activity*
of the plug-in.

For example, if we want to jump to the *MainActivity* of the plug-in, the
code is as follows:

```
Intent intent = new Intent("jianqiang.com.hostapp.
VIEW");
intent.putExtra(ProxyActivity.EXTRA_DEX_PATH,
mPluginItems.get(position).pluginPath);
intent.putExtra(ProxyActivity.EXTRA_CLASS,
mPluginItems.get(position).packageInfo.packageName +
".MainActivity");
startActivity(intent);
```

The above intent carries two parameters:

1) *EXTRA_DEX_PATH*, the path of the *dex* in the plug-in.

2) *EXTRA_CLASS*, the full name of the *Activity* in the plug-in (*packageName + className*).

252 ■ Android App-Hook and Plug-In Technology

9.1.2.2 Communication between ProxyActivity *and Plug-In* Activity

To simplify the logic of *ProxyActivity*, I abstract the logic of creating a *ClassLoader* for the plug-in and loading the resources of the plug-in into a parent class *BaseHostActivity*. We introduced this technique in Chapter 7.

Let's focus on the special logic of *ProxyActivity*.

In the method *onCreate()* of *ProxyActivity*, we get an instance of the *Activity* from the plug-in; its name is *mRemoteActivity*. *mRemoteActivity* can help us do the following:

1) Call the method *setProxy()* of *mRemoteActivity* and pass "this" (instance of *ProxyActivity*) and the path of the plug-in to the *Activity* of the plug-in.

```
//Create the instance of Activity using reflection
Class<?> localClass = dexClassLoader.
loadClass(className);
Constructor<?> localConstructor = localClass.
getConstructor(new Class[] {});
Object instance = localConstructor.newInstance(new
Object[] {});
mRemoteActivity = (Activity) instance;

//Execute the setProxy method of the Plug-In Activity
to establish a bidirectional reference
RefInvoke.invokeInstanceMethod(instance, "setProxy",
  new Class[] { Activity.class, String.class },
  new Object[] { this, mDexPath });
```

Now *ProxyActivity* has a reference to the *Activity* of the plug-in app, and the *Activity* of the plug-in app also has a reference to *ProxyActivity*. Later we will talk about how to use the variable "this" passed from the method *setProxy()* in the *Activity* of the plug-in.

2) In the method *onCreate()* of *ProxyActivity*, we invoke the method *onCreate()* of *mRemoteActivity* by reflection. *ProxyActivity* of the HostApp is the old artist, and *mRemoteActivity* of plug-in is the puppet.

We implement the method *onResume()* as follows:

```
@Override
protected void onResume() {
    super.onResume();
```

```
try {
    Method method = localClass.
    getDeclaredMethod(methodName, new Class[]
    { });
    method.setAccessible(true);
    method.invoke(mRemoteActivity, new
    Object[] { });
} catch (Exception e) {
e.printStackTrace();
}
}
```

However, there is a performance issue in the code above. The method *onResume()* will be invoked multiple times, and the reflection will be invoked many times. To optimize performance, we store the *method* object in a dictionary. When we want to invoke the method *onResume()*, we fetch the *method* object from the dictionary, and then call the *invoke()* method of *method* object.

There're many lifecycle methods of *Activity*, such as *onCreate()*, *onActivityResult()*, *onRestart()*, *onStart()*, *onPause()*, *onStop()*, and *onDestroy()*. All the methods can follow route to improve performance. We encapsulate this logic into the method *instantiateLifecircleMethods()*, as follows:

```
protected void instantiateLifecircleMethods(Class<?>
localClass) {
    String[] methodNames = new String[] {
        "onRestart",
        "onStart",
        "onResume",
        "onPause",
        "onStop",
        "onDestroy"
    };
    for (String methodName : methodNames) {
      Method method = null;
      try {
        method = localClass.getDeclaredMethod
        (methodName, new Class[] { });
        method.setAccessible(true);
      } catch (NoSuchMethodException e) {
        e.printStackTrace();
      }
```

```
    mActivityLifecircleMethods.put(methodName, method);
    }

    Method onCreate = null;
    try {
        onCreate = localClass.getDeclaredMethod
        ("onCreate", new Class[] { Bundle.class });
        onCreate.setAccessible(true);
    } catch (NoSuchMethodException e) {
        e.printStackTrace();
    }
    mActivityLifecircleMethods.put("onCreate",
    onCreate);

    Method onActivityResult = null;
    try {
        onActivityResult = localClass.getDeclaredMeth
        od("onActivityResult",
            new Class[] { int.class, int.class,
            Intent.class });
        onActivityResult.setAccessible(true);
    } catch (NoSuchMethodException e) {
        e.printStackTrace();
    }
    mActivityLifecircleMethods.put("onActivityResult",
    onActivityResult);
}
```

And then, we can implement the method *onResume()* as follows:

```
@Override
  protected void onResume() {
    super.onResume();
    Method onResume = mActivityLifecircleMethods.
    get("onResume");
    if (onResume != null) {
      try {
          onResume.invoke(mRemoteActivity, new
          Object[] { });
      } catch (Exception e) {
          e.printStackTrace();
      }
    }
  }
```

9.1.2.3 The Logic of Activity in the Plug-In

We encapsulate some common code from the plug-in into *BasePluginActivity.*

As mentioned earlier, the *ProxyActivity* of the HostApp will pass itself ("this") to the variable "that" of *Activity* in the plug-in. The variable "that" is defined in *BasePluginActivity.*

So, what is the use of "that"? In other words, the *Activity* of the plug-in is only a puppet without life. So, when we write the following code in the *MainActivity* of the plug-in, it will throw an exception during runtime:

```
@Override
protected void onCreate(Bundle savedInstanceState) {
        this.setContentView(R.layout.activity_main);
        this.findViewById(R.id.button1);
}
```

The keyword "this" points to the current object, and it is an *Activity* of the *apk*, but the *Activity* of the plug-in is not a real *Activity*, which means we can't use the keyword "this" in the *Activity* of the plug-in any more.

We can use the keyword "that" instead to avoid this runtime exception:

```
that.setContentView(R.layout.activity_main);
that.findViewById(R.id.button1);
```

9.1.3 Jump in the Plug-In*

In Plugin1, jumping from *MainActivity* to *SecondActivity* is actually a navigation between two *ProxyActivities* of the HostApp.

We change some code in the plug-in:

1) Add the logic for jumping from *MainActivity* to *SecondActivity*:

```
public class MainActivity extends BasePluginActivity {

  private static final String TAG =
  "Client-MainActivity";

  @Override
  protected void onCreate(Bundle savedInstanceState) {
```

---

* Sample code: https://github.com/BaoBaoJianqiang/That1.1

```
that.setContentView(R.layout.activity_main);

//startActivity, Jump inside the plugin
Button button1 = (Button) that.findViewById
(R.id.button1);
button1.setOnClickListener(new OnClickListener() {
  @Override
  public void onClick(View v) {
    Intent intent = new Intent(AppConstants.
    PROXY_VIEW_ACTION);
    intent.putExtra(AppConstants.EXTRA_DEX_PATH,
    dexPath);
    intent.putExtra(AppConstants.EXTRA_CLASS,
    "jianqiang.com.plugin1.SecondActivity");
    that.startActivity(intent);
  }
```

2) The logic in *SecondActivity*:

```
public class SecondActivity extends BasePluginActivity {

  private static final String TAG =
  "Client-SecondActivity";

  @Override
protected void onCreate(Bundle savedInstanceState) {
    that.setContentView(R.layout.second);
  }
}
```

Maybe we are not used to this code style. We can find the keyword "that" everywhere, rather than "this." Where *super.onCreate()* is present, we will try to solve these inconveniences in the next section.

9.1.4 Eliminate the Keyword "that"*

In the "That" framework, the lifecycle method of an *Activity* in the plug-in needn't call its parent method, because it doesn't have its own lifecycle.

---

* Sample code: https://github.com/BaoBaoJianqiang/That1.2

For example, the method *onCreate()*, if we write the code as follows, will throw an exception at runtime.

```
@Override
protected void onCreate(Bundle savedInstanceState) {
    super.onCreate(savedInstanceState);

    that.setContentView(R.layout.activity_main);
    that.findViewById(R.id.button1);
}
```

Therefore, we can't use the keyword "super" anymore.

In addition, the keyword "that" is widely used in plug-ins, such as the methods *setContentView()*, *findViewById()*. In fact, the complete statement is as follows:

```
this.that.setContentView(R.layout.activity_main);
this.that.findViewById(R.id.button1);
```

So if we don't use "that," it will throw a compile error when we invoke the method *setContentView()* in the plug-in directly, as it will invoke the lifecycle methods of an *Activity* in the plug-in, but actually, *Activity* in the plug-in doesn't have a lifecycle.

We don't want to use "that" frequently, so we try to solve this syntax problem through object-oriented programming.

Let's come back to the *BasePluginActivity* of the plug-in to add some empty methods to it, such as *setContentView()*, *findViewById()*, and *startActivity()*:

```
public class BasePluginActivity extends Activity {
  @Override
  protected void onCreate(Bundle savedInstanceState) {
  }

  @Override
  public void setContentView(int layoutResID) {
    that.setContentView(layoutResID);
  }

  @Override
  public View findViewById(int id) {
    return that.findViewById(id);
  }
```

```
@Override
public void startActivity(Intent intent) {
  that.startActivity(intent);
}
}
```

Well, in the *MainActivity* of the plug-in, we can write code normally as before, the keyword "that" has gone, and the keyword "super" is back:

```
public class MainActivity extends BasePluginActivity {

  private static final String TAG = "Client-MainActivity";

  @Override
  protected void onCreate(Bundle savedInstanceState) {
    super.onCreate(savedInstanceState);

    setContentView(R.layout.activity_main);

    //startActivity, Jump inside the plugin
    Button button1 = (Button) findViewById
    (R.id.button1);
    button1.setOnClickListener(new OnClickListener() {
      @Override
      public void onClick(View v) {
        Intent intent = new Intent(AppConstants.
        PROXY_VIEW_ACTION);
        intent.putExtra(AppConstants.EXTRA_DEX_PATH,
        dexPath);
        intent.putExtra(AppConstants.EXTRA_CLASS,
        "jianqiang.com.plugin1.SecondActivity");
        startActivity(intent);
      }
    });
  }
}
```

The methods *setContentView()*, *findViewById()*, *startActivity()* in the code above, will call the same method of *BasePluginActivity*, and the method *super.onCreate()* is also to call the method *onCreate()* of *BasePluginActivity*.

We can't eliminate the keyword "that" at all, for example, the method *setResult()* of *Activity*.

The method *setResult* defined in the *Activity* has a modifier *final*, so this method can't be overridden; we have to use *that.setResult()*.

In the *Activity*, there are many methods with the modifier *final*.

9.1.5 Jump Out*

We hope to improve the jumping mechanism in the plug-in; for example, to jump from the plug-in to the HostApp or to another plug-in.

9.1.5.1 Preparation for Jumping Out

Let's do some preparations first.

1) Add another plug-in project named Plugin2.

When the app navigates from Plugin1 to Plugin2, Plugin1 needs to know the *dexPath* (the path of the *dex* file) of Plugin2. So it's time to create a class *MyPlugins*, which is a container. The *dexPath* of each plug-in app is defined in this class, shown as follows:

```
public class MyPlugins {
  public final static HashMap<String, String> plug-ins
  = new HashMap<String, String>();
}
```

2) When the *MainActivity* of the HostApp resolves Plugin1 and Plugin2, put the *dexPath* of Plugin1 and Plugin2 into *MyPlugins*:

```
File file1 = getFileStreamPath("plugin1.apk");
File file2 = getFileStreamPath("plugin2.apk");
File[] plugins = {file1, file2};

for (File plugin : plugins) {
  PluginItem item = new PluginItem();
  item.pluginPath = plugin.getAbsolutePath();
  item.packageInfo = DLUtils.getPackageInfo(this,
  item.pluginPath);
  mPluginItems.add(item);
  MyPlugins.plugins.put(plugin.getName(), item.
  pluginPath);
}
```

---

* Sample code: https://github.com/BaoBaoJianqiang/That1.3

9.1.5.2 Jump to Another Plug-In

This example illustrates how to jump from *MainActivity* of Plugin2 to *SecondActivity* of Plugin1; actually it is a jump between two *ProxyActivities* of the HostApp, as follows:

```java
public class MainActivity extends BasePluginActivity {

  @Override
  protected void onCreate(Bundle savedInstanceState) {
    super.onCreate(savedInstanceState);
    setContentView(R.layout.activity_main);

    //startActivity, Plug-In jump
    Button button1 = (Button) findViewById
    (R.id.button1);
    button1.setOnClickListener(new View.
    OnClickListener() {
      @Override
      public void onClick(View v) {
        String plugin1DexPath = MyPlugins.plugins.
        get("plugin1.apk");

        Intent intent = new Intent(AppConstants.
        PROXY_VIEW_ACTION);
        intent.putExtra(AppConstants.EXTRA_DEX_PATH,
        plugin1DexPath);
        intent.putExtra(AppConstants.EXTRA_CLASS,
        "jianqiang.com.plugin1.SecondActivity");
        startActivity(intent);
      }
    });
  }
}
```

9.1.5.3 Jump to the HostApp

This example demonstrates how to jump from *MainActivity* of Plug1 to *MainActivity* of the HostApp:

```java
Button button3 = (Button) findViewById(R.id.button3);
    button3.setOnClickListener(new OnClickListener() {
      @Override
      public void onClick(View v) {
```

```
    Intent intent = new Intent();
    intent.putExtra("userName", "baojianqiang");
    ComponentName componentName = new
    ComponentName("jianqiang.com.hostapp",
    "jianqiang.com.hostapp.MainActivity");
    intent.setComponent(componentName);
    startActivity(intent);
  }
});
```

9.1.6 Use Interface-Oriented Programming in *Static-Proxy**

Now we focus on the *ProxyActivity* of the HostApp. We use reflection to execute the methods *onRestart(), onStart(), onResume()* and so on. The syntax of reflection is ugly, shown as follows:

```
public class ProxyActivity extends BaseHostActivity {

  private static final String TAG = "ProxyActivity";

  private String mClass;

  private Activity mRemoteActivity;
  private HashMap<String, Method>
  mActivityLifecircleMethods = new HashMap<String,
  Method>();

  @Override
  protected void onCreate(Bundle savedInstanceState) {
    super.onCreate(savedInstanceState);
    mDexPath = getIntent().getStringExtra
    (AppConstants.EXTRA_DEX_PATH);
    mClass = getIntent().getStringExtra(AppConstants.
    EXTRA_CLASS);

    loadClassLoader();
    loadResources();

    launchTargetActivity(mClass);
  }

  void launchTargetActivity(final String className) {
```

---

* Sample code: https://github.com/BaoBaoJianqiang/That1.4

```
try {
    // Reflects the plugin's Activity object
    Class<?> localClass = dexClassLoader.
    loadClass(className);
    Constructor<?> localConstructor = localClass.
    getConstructor(new Class[] {});
    Object instance = localConstructor.
    newInstance(new Object[] {});
    mRemoteActivity = (Activity) instance;

    //Execute the setProxy method of the Plug-In
    Activity to establish a bidirectional
    reference
    Method setProxy = localClass.
    getMethod("setProxy", new Class[] { Activity.
    class, String.class });
    setProxy.setAccessible(true);
    setProxy.invoke(instance, new Object[] {
    this, mDexPath });

    //One-time reflection activity life cycle
    function
    instantiateLifecircleMethods(localClass);

    //Execute the onCreate method of the Plug-In
    Activity
    Method onCreate = mActivityLifecircleMethods.
    get("onCreate");
    Bundle bundle = new Bundle();
    onCreate.invoke(instance, new Object[] {
    bundle });
} catch (Exception e) {
    e.printStackTrace();
}
}

protected void instantiateLifecircleMethods(Class<?>
localClass) {
    String[] methodNames = new String[] {
        "onRestart",
        "onStart",
```

```java
        "onResume",
        "onPause",
        "onStop",
        "onDestroy"
    };
    for (String methodName : methodNames) {
      Method method = null;
      try {
            method = localClass.getDeclaredMethod
            (methodName, new Class[] { });
            method.setAccessible(true);
      } catch (NoSuchMethodException e) {
            e.printStackTrace();
      }
      mActivityLifecircleMethods.put(methodName, method);
Method onCreate = null;
  try {
        onCreate = localClass.getDeclaredMethod
        ("onCreate", new Class[] { Bundle.class });
        onCreate.setAccessible(true);
  } catch (NoSuchMethodException e) {
        e.printStackTrace();
  }
  mActivityLifecircleMethods.put("onCreate",
  onCreate);

  Method onActivityResult = null;
  try {
        onActivityResult = localClass.getDeclaredMeth
        od("onActivityResult",
          new Class[] { int.class, int.class,
          Intent.class });
        onActivityResult.setAccessible(true);
  } catch (NoSuchMethodException e) {
        e.printStackTrace();
  }
  mActivityLifecircleMethods.put("onActivityResult",
  onActivityResult);
}
```

```
@Override
protected void onStart() {
  super.onStart();
  Method onStart = mActivityLifecircleMethods.
  get("onStart");
  if (onStart != null) {
    try {
        onStart.invoke(mRemoteActivity, new
        Object[] {});
    } catch (Exception e) {
        e.printStackTrace();
    }
  }
}
}
```

We introduced interface-oriented programming in the previous section; it's widely used in plug-in programming.

Let's design an interface *IRemoteActivity*:

```
public interface IRemoteActivity {
  public void onStart();
  public void onRestart();
  public void onActivityResult(int requestCode, int
  resultCode, Intent data);
  public void onResume();
  public void onPause();
  public void onStop();
  public void onDestroy();
  public void onCreate(Bundle savedInstanceState);
  public void setProxy(Activity proxyActivity, String
  dexPath);
}
```

Then we make *BasePluginActivity* implement the interface *IRemoteActivity*:

```
public class BasePluginActivity extends Activity
implements IRemoteActivity {
```

Originally the methods *onCreate()*, *onResume()* and others have a modifiler *protected*, now we need set it to *public*.

Then we can use interface-oriented programming in *ProxyActivity* of the HostApp. The new implementation of *ProxyActivity* is as follows:

```
public class ProxyActivity extends BaseHostActivity {

  private static final String TAG = "ProxyActivity";

  private String mClass;

  private IRemoteActivity mRemoteActivity;
  private HashMap<String, Method>
  mActivityLifecircleMethods = new HashMap<String,
  Method>();

  @Override
  protected void onCreate(Bundle savedInstanceState) {
    super.onCreate(savedInstanceState);
    mDexPath = getIntent().getStringExtra(AppConstants.
    EXTRA_DEX_PATH);
    mClass = getIntent().getStringExtra(AppConstants.
    EXTRA_CLASS);

    loadClassLoader();
    loadResources();

    launchTargetActivity(mClass);
  }

  void launchTargetActivity(final String className) {
    try {
        //Activity objects that reflect the plugin
        Class<?> localClass = dexClassLoader.
        loadClass
        (className);
        Constructor<?> localConstructor = localClass.
        getConstructor(new Class[] {});
        Object instance = localConstructor.
        newInstance(new Object[] {});

        mRemoteActivity = (IRemoteActivity) instance;
        mRemoteActivity.setProxy(this, mDexPath);
```

```
        //Execute the onCreate method of the Plug-In
        Activity
        Bundle bundle = new Bundle();
        mRemoteActivity.onCreate(bundle);
    } catch (Exception e) {
        e.printStackTrace();
    }
}

@Override
protected void onActivityResult(int requestCode, int
resultCode, Intent data) {
    Log.d(TAG, "onActivityResult resultCode=" +
    resultCode);
    mRemoteActivity.onActivityResult(requestCode,
    resultCode, data);
    super.onActivityResult(requestCode, resultCode,
    data);
}

@Override
protected void onStart() {
    super.onStart();
    mRemoteActivity.onStart();
}

@Override
protected void onRestart() {
    super.onRestart();
    mRemoteActivity.onRestart();
}
@Override
protected void onResume() {
    super.onResume();
    mRemoteActivity.onResume();
}

@Override
protected void onPause() {
    super.onPause();
    mRemoteActivity.onPause();
}
```

```
@Override
protected void onStop() {
  super.onStop();
  mRemoteActivity.onStop();
}

@Override
protected void onDestroy() {
  super.onDestroy();
  mRemoteActivity.onDestroy();
}
}
```

Is it clearer now? This is the power of interface-oriented programming. Next, we should add more lifecycle methods for *Activity* in *IRemoteActivity*, such as:

- *public void onSaveInstanceState(Bundle outState)*;

- *public void onNewIntent(Intent intent)*;

- *public void onRestoreInstanceState(Bundle savedInstanceState)*;

- *public boolean onTouchEvent(MotionEvent event)*;

- *public boolean onKeyUp(int keyCode, KeyEvent event)*;

- *public void onWindowAttributesChanged(LayoutParams params)*;

- *public void onWindowFocusChanged(boolean hasFocus)*;

9.1.7 Support for *LaunchMode**

Now let's discuss how to support *LaunchMode* in the "That" framework. It was fixed by Tao Zhang.

The content of this section is based on Tao Zhang's framework *CJFrameworkForAndroid*[†].

Let's have a quick look at *LaunchMode* first.

9.1.7.1 Overview of LaunchMode

1) *Launchmode* is declared in the *AndroidManifest.xml* as follows:

```
<activity android:launchMode="singleTask"></activity>
```

---

* Sample code: https://github.com/BaoBaoJianqiang/That1.5
† Tao Zhang, CJFrameforAndroid. https://github.com/kymjs/CJFrameForAndroid. (Accessed 24 June 2019.)

2) *LaunchMode* has four values

Android apps have an *Activity* stack, and each time the app launches an *Activity* it will put this *Activity* on top of the stack. According to the different usage of the stack, *LaunchMode* has four values, *standard*, *singleTop*, *singleTask*, and *singleInstance*, shown as follows:

- *standard*, the default mode. Each time the app launches an *Activity*, it will create a new instance of this *Activity*. Even if the app jumps from *Activity1* to *Activity1*, because the *LaunchMode* of *Activity1* is *standard* there are two instances of *Activity1* in the stack.

Activity is widely used in apps. There are hundreds of *Activities* in the app. Almost all the *Activities* have the same *LaunchMode*: *standard*.

Let's introduce another scenario based on *LaunchMode*: Push. A news app always sends push messages to the users; when the user clicks on the push message, the app will navigate to a *DetailActivity* to show the detail of the news. If we set the *LaunchMode* of *DetailActivity* to *standard*, when the user receives ten push messages from this news app he clicks each message and opens *DetailActivity* ten times to see the different news items. But he also needs to click ten times to come back to the original *Activity*.

Obviously, *standard* is not suitable for this scenario. We want to click once to get back to the original *Activity*.

- *singleTop*. If the current page is *Activity1* (at the top of the stack), let's navigate to *Activity1*, because the *LaunchMode* of *Activity1* is *singleTop*, the instance of *Activity1* at the top of the stack will be used directly, instead of creating a new instance.

  We will find that *singleTop* is suitable for the push message.

- *singleTask*. If we want to open *Activity1*, and there is an instance of *Activity1* in the stack, because the *LaunchMode* of *Activity1* is *singleTask*, this instance will be used. At the same time, all the instances in the stack, from this instance to the top *Activity*, will be destroyed, which means that this instance will be the top of the *Activity* stack.

  singleTask is generally used to jump back to the home page of the app. The instance of the home page is only one.

- *singleInstance.* If the *LaunchMode* of the *Activity* has this value, the Android system will create an instance of this *Activity* in a new stack, and all the third app can visit this instance in the stack.

It's difficult to understand. Let's take the camera as an example. My App wants to use the camera; the camera is a system app in Android. My app and camera are two apps running in different processes. My app will navigate the *Activity* of the camera. The *LaunchMode* of this *Activity* is *singleInstance.* After my app finishes using the camera, it goes back to my app. Because the other apps maybe want to use the camera too.

9.1.7.2 Plug-In Solutions for LaunchMode

We have introduced the basic concept of *LaunchMode.* Now let's study how to support it in the plug-in. The solution is to simulate the function of the physical back button.

First, write a *Singleton CJBackStack*, with a field *atyStack*; *atyStack* is a collection to store all the open *Activities* of the plug-in.

Let's have a look at the method *launch()* of *CJBackStack.*

```
public void launch(IRemoteActivity pluginAty) {
  atyStack.add(pluginAty);

  if (atyStack.size() == 1)
     return;

  if(pluginAty.getLaunchMode() == LaunchMode.STANDARD)
     return;

  if(pluginAty.getLaunchMode() == LaunchMode.
  SINGLETOP) {
    //Countdown to the second element
    int index = atyStack.size() - 2;
    if (atyStack.get(index).getClass().getName().equals(
       pluginAty.getClass().getName())) {
     remove(atyStack.size() - 2);
    }
  }

  for (int i = atyStack.size() - 2; i >= 0; i--) {
```

```
    if (atyStack.get(i).getClass().getName().
    equals(pluginAty.getClass().getName())) {
      switch (pluginAty.getLaunchMode()) {
        case LaunchMode.SINGLETASK: // Stack unique
          // Here, since each remove() and atyStack.
          size() will decrease, the third paragraph of
          the for statement is omitted.
          for (int j = i; j < atyStack.size() - 1;) {
            remove(j);
          }
          break;
        case LaunchMode.SINGLEINSTANCE://Unique in
        application
          remove(i);
          break;
      }
    }
  }
}
```

Each time we launch an *Activity* from the plug-in, we put the *Activity* in the *atyStack*. There are five branches.

1) If there is only one *Activity* in *atyStack*, do nothing. *LaunchMode* does not work at this time.

2) If the *LaunchMode* of the current *Activity* is *standard*, do nothing.

3) If the *LaunchMode* of the current *Activity* is *singleTop*, then check the second top element in *atyStack*. If the type of this element is the same as the top *Activity*, let's remove the second top element from *atyStack* and execute its method *finish()* to close this *Activity*. When we press the backpress button, the app will jump back to the third top element, which simulate the scenario *singleTop*.

```
private void remove(int index) {
    IRemoteActivity aty = atyStack.get(index);
    atyStack.remove(index);
    if (aty instanceof BasePluginActivity) {
      ((BasePluginActivity) aty).finish();
    }
  }
```

```
public void finish(IRemoteActivity aty) {
  for (int i = atyStack.size() - 2; i >= 0; i--) {
    if (aty.equals(atyStack.get(i))) {
      remove(i);
    }
  }
}
```

4) When the *LaunchMode* of the current *Activity* is *singleTask*, it is necessary to traverse the other elements of *atyStack* to see if there's an element with the same type as the current *Activity*. If yes, remove all the elements between this element and the current *Activity* from *atyStack* and execute the method *finish()* to close them.

5) When the *LaunchMode* of the current *Activity* is *singleInstance*, it is necessary to traverse the other elements of *atyStack* to see if there's an element with same type as the current *Activity*. If yes, remove it and execute its method *finish()* to close it.

In this demo, in the *MainActivity* of Plugin2, let's click the button *singleTop* to jump to *ActivityA*. The *LaunchMode* of *ActivityA* is *singleTop*.

There is a button in *ActivityA*, let's click this button to jump to *ActivityA* again. Then let's press the backpress button, we find that the app goes directly to the *MainActivity* of Plugin2. This means that *singleTop* is simulated successfully in the plug-in.

This section introduces the "That" framework of *Activity*. We don't need more knowledge of the Android system; we only need to master the proxy pattern.

The advantage of the "That" framework is that we don't need to hook the internal API of the Android system.

9.2 THE PLUG-IN SOLUTION FOR *SERVICE* AND *BROADCASTRECEIVER* BASED ON *STATIC-PROXY*

This section continues to introduce how to support *Service* and *BroadcastReceiver* in the "That" framework step by step.

9.2.1 *Static-Proxy* in *Service**

In the "That" framework, we use to *ProxyActivity* to control all the *Activities* from the plug-in. This solution is also suitable for *Service*. We try

---

* Sample code: https://github.com/BaoBaoJianqiang/That3.1

to write a demo, That3.1, based on the demo That1.5 introduced in Section 9.1.7. There are three core classes; the following table lists all mapping of these three classes between the plug-in solutions for *Activity* and *Service*:

Activity	Service
ProxyActivity	ProxyService
BasePluginActivity	BasePluginService
IRemoteActivity	IRemoteService

1) *IRemoteService*, the interface needs to override the following five methods of *Service*; it is much simpler than *Activity* with a lot of lifecycle methods.

- *onCreate*

- *onStartCommand*

- *onDestroy*

- *onBind*

- *onUnbind*

2) *BasePluginService*, the parent class of *Service* in the plug-in, acts as a puppet like *BasePluginActivity*.

```
public class BasePluginService extends Service
implements IRemoteService {

  public static final String TAG = "DLBasePluginService";
  private Service that;
  private String dexPath;

  @Override
  public void setProxy(Service proxyService, String
  dexPath) {
    that = proxyService;
    this.dexPath = dexPath;
  }

  @Override
  public void onCreate() {
    Log.d(TAG, TAG + " onCreate");
  }
```

```java
  @Override
  public int onStartCommand(Intent intent, int flags,
  int startId) {
    Log.d(TAG, TAG + " onStartCommand");
    return 0;
  }

  @Override
  public void onDestroy() {
    Log.d(TAG, TAG + " onDestroy");
  }

  @Override
  public IBinder onBind(Intent intent) {
    Log.d(TAG, TAG + " onBind");
    return null;
  }

  @Override
  public boolean onUnbind(Intent intent) {
    Log.d(TAG, TAG + " onUnbind");
    return false;
  }
}
```

3) *ProxyService*, this class acts as an artist manipulating the puppets.

```java
public class ProxyService extends Service {

  private static final String TAG = "DLProxyService";

  private String mClass;
  private IRemoteService mRemoteService;

  @Override
  public void onCreate() {
    super.onCreate();
  }

  @Override
  public int onStartCommand(Intent intent, int flags,
  int startId) {
    super.onStartCommand(intent, flags, startId);
```

```
mDexPath = intent.getStringExtra(AppConstants.
EXTRA_DEX_PATH);
mClass = intent.getStringExtra(AppConstants.
EXTRA_CLASS);

loadClassLoader();

try {
    //get Activity objects of Plug-In by
    reflection
    Class<?> localClass = dexClassLoader.
    loadClass(mClass);
    Constructor<?> localConstructor = localClass.
    getConstructor(new Class[] {});
    Object instance = localConstructor.
    newInstance(new Object[] {});

    mRemoteService = (IRemoteService) instance;
    mRemoteService.setProxy(this, mDexPath);

    return mRemoteService.onStartCommand(intent,
    flags, startId);
} catch (Exception e) {
    e.printStackTrace();
    return 0;
}
}

@Override
public void onDestroy() {
  super.onDestroy();
  Log.d(TAG, TAG + " onDestroy");

  mRemoteService.onDestroy();
}

@Override
public IBinder onBind(Intent intent) {
  Log.d(TAG, TAG + " onBind");
  return mRemoteService.onBind(intent);
}
```

```
@Override
public boolean onUnbind(Intent intent) {
  Log.d(TAG, TAG + " onUnbind");
  return mRemoteService.onUnbind(intent);
}

protected String mDexPath;
protected ClassLoader dexClassLoader;

protected void loadClassLoader() {
  File dexOutputDir = this.getDir("dex", Context.
  MODE_PRIVATE);
  final String dexOutputPath = dexOutputDir.
  getAbsolutePath();
  dexClassLoader = new DexClassLoader(mDexPath,
      dexOutputPath, null, getClassLoader());
  }
}
```

Now let's write a class *TestService1* in Plugin2 and write a class *TestService2* in Plugin2; they both inherit *BasePluginService*. Let's take *TestService1* as an example:

```
public class TestService1 extends BasePluginService {

  private static final String TAG = "TestService1";

  @Override
  public void onCreate() {
    super.onCreate();
    Log.e(TAG, "onCreate");
  }

  @Override
  public int onStartCommand(Intent intent, int flags,
  int startId) {
    Log.e(TAG, "onStartCommand");
    return super.onStartCommand(intent, flags,
    startId);
  }
```

```
@Override
public void onDestroy() {
  super.onDestroy();
  Log.d(TAG, TAG + " onDestroy");
  }
}
```

4) In the HostApp, we invoke the *TestService* of Plugin1 as follows:

```
public void startService1InPlugin1(View view) {
  Intent intent = new Intent();
  intent.setClass(this, ProxyService.class);
  intent.putExtra(AppConstants.EXTRA_DEX_PATH,
  pluginItem1.pluginPath);
  intent.putExtra(AppConstants.EXTRA_CLASS,
  pluginItem1.packageInfo.packageName +
  ".TestService1");

  startService(intent);
  }

public void stopService1InPlugin1(View view) {
  Intent intent = new Intent();
  intent.setClass(this, ProxyService.class);
  intent.putExtra(AppConstants.EXTRA_DEX_PATH,
  pluginItem1.pluginPath);
  intent.putExtra(AppConstants.EXTRA_CLASS,
  pluginItem1.packageInfo.packageName +
  ".TestService1");

  stopService(intent);
  }
```

Now, a simple plug-in demo to support *Service* is completed, but there are three issues:

9.2.1.1 Issue 1

Service is different from *Activity*. When we open the same *Activity* many times, the Android system will create multiple instances of this *Activity*, so in the "That" framework, one *ProxyActivity* corresponds to multiple *Activities* of the plug-in.

But one *Service* only corresponds to one instance, so one *ProxyService* can't correspond to the multiple *Services* of the plug-in.

In the demo above, when we start *TestService1* of Plugin1 or *TestService2* of Plugin2, we use the same instance of *ProxyService*.

```
public void startService2InPlugin2(View view) {
    Intent intent = new Intent();
    intent.setClass(this, ProxyService.class);
    intent.putExtra(AppConstants.EXTRA_DEX_PATH,
    pluginItem2.pluginPath);
    intent.putExtra(AppConstants.EXTRA_CLASS,
    pluginItem2.packageInfo.packageName +
    ".TestService2");

    startService(intent);
}

public void stopService2InPlugin2(View view) {
    Intent intent = new Intent();
    intent.setClass(this, ProxyService.class);
    intent.putExtra(AppConstants.EXTRA_DEX_PATH,
    pluginItem2.pluginPath);
    intent.putExtra(AppConstants.EXTRA_CLASS,
    pluginItem2.packageInfo.packageName +
    ".TestService2");

    stopService(intent);
}
```

When I click the button "start TestService1" and "start TestService2" step by step, and then click the button "stop TestService1," we find that *TestService1* and *TestService2* stop at the same time.

9.2.1.2 Issue 2
When I write the method *bindService()* and *unbindService()*, I find that the variable *mRemoteService* is *null* in the method *binding()* of *BasePluginService*, that is why it crashed:

```
public void bindService3InPlugin1(View view) {
    Intent intent = new Intent();
    intent.setClass(this, ProxyService.class);
```

```
    intent.putExtra(AppConstants.EXTRA_DEX_PATH,
    pluginItem1.pluginPath);
    intent.putExtra(AppConstants.EXTRA_CLASS,
    pluginItem1.packageInfo.packageName +
    ".TestService3");

    bindService(intent, mConnection, Context.
    BIND_AUTO_CREATE);
}

public void unbindService3InPlugin1(View view) {
    unbindService(mConnection);
}
```

The variable *mRemoteService* is only instantiated in the method *onStartCommand()* of *ProxyService*. But the method *bind()* of *Service* doesn't invoke the method *onStartCommand()*. We can find the truth in Figure 9.3.

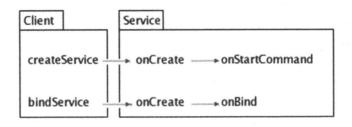

FIGURE 9.3   Two paths of *Service*.

9.2.1.3 Issue 3
When we debug the code in the plug-in, we find the method *onCreate()* of *Service* has never been invoked.

Because *Service* in the plug-in is a normal class, and it doesn't have a lifecycle, including *onCreate()*.

9.2.2 Support *bindService()**
In this section, we will solve issue 2 and issue 3.

To solve issue 2 that the variable *mRemoteService* is *null* when the method *onBind()* of *ProxyService* is executed, we need copy the code in the

---

* For the example code of this section, please refer to https://github.com/Baobaojianqiang/That3.2

method *onStartCommand()* of *ProxyService* and paste it into the method *onBind()*.

To solve issue 3 that the method *onCreate()* of *ProxyService* is not executed, we need to execute the methods *onStartCommand()* and *onBind()* of *ProxyService* manually.

The code is as follows:

```
@Override
public int onStartCommand(Intent intent, int flags,
int startId) {
  super.onStartCommand(intent, flags, startId);

  mDexPath = intent.getStringExtra(AppConstants.
  EXTRA_DEX_PATH);
  mClass = intent.getStringExtra(AppConstants.
  EXTRA_CLASS);

  loadClassLoader();

  try {
      //get Server objects of Plug-In by reflection
      Class<?> localClass = dexClassLoader.
      loadClass(mClass);
      Constructor<?> localConstructor = localClass.
      getConstructor(new Class[] {});
      Object instance = localConstructor.
      newInstance(new Object[] {});

      mRemoteService = (IRemoteService) instance;
      mRemoteService.setProxy(this, mDexPath);
        mRemoteService.onCreate();
      return mRemoteService.onStartCommand(intent,
      flags, startId);
  } catch (Exception e) {
      e.printStackTrace();
      return 0;
  }
}

@Override
public IBinder onBind(Intent intent) {
  Log.d(TAG, TAG + " onBind");
```

```
mDexPath = intent.getStringExtra(AppConstants.
EXTRA_DEX_PATH);
mClass = intent.getStringExtra(AppConstants.
EXTRA_CLASS);

loadClassLoader();

try {
    //get Server objects of Plug-In by reflection
    Class<?> localClass = dexClassLoader.
    loadClass(mClass);
    Constructor<?> localConstructor = localClass.
    getConstructor(new Class[] {});
    Object instance = localConstructor.
    newInstance(new Object[] {});

    mRemoteService = (IRemoteService) instance;
    mRemoteService.setProxy(this, mDexPath);
      mRemoteService.onCreate();
    return mRemoteService.onBind(intent);
} catch (Exception e) {
    e.printStackTrace();
    return null;
}
}
```

There is a performance issue in the above code: each time the method *onBind()* is invoked, we should create a *ClassLoader* from the plug-in. In fact, we can create all the *ClassLoaders* of the plug-ins at one time, and store them in a *HashMap*, which we can take out whenever we want to use it.

9.2.3 *StubService**

For most apps, there are a lot Activities but few Services in an app. The number of Services in one app is less than ten.

Based on this fact, we can pre-write ten *ProxyServices* in the HostApp, and name them from *ProxyService1* to *ProxyService10*.

The next step is to create a strict mapping between each *ProxyService* and real *Service* of the plug-in; for example, *ProxyService1* corresponds to

---

* For the example code of this section, please refer to https://github.com/Baobaojianqiang/That3.3

the *TestService1* of the plug-in, *ProxyService2* corresponds to *TestService2* of the plug-in, and so on. We can maintain a *JSON* string to record this mapping, shown as follows:

```
{
  pluginServices: [
    {
      proxy: jianqiang.com.hostapp.ProxyService1,
      realService: jianqiang.com.plugin1.TestService1
    },
    {
      proxy: jianqiang.com.hostapp.ProxyService2,
      realService: jianqiang.com.plugin2.TestService2
    },
    {
      proxy: jianqiang.com.hostapp.ProxyService3,
      realService: jianqiang.com.plugin1.TestService3
    }
  ]
}
```

We can download this *JSON* string from the remote server. We can also store this *JSON* string in the plug-in app.

Take the *TestService1* of Plugin1 as an example:

1) Design a *Singleton ProxyServiceManager*, to obtain the corresponding *ProxyService* according to the *Service* of the plug-in.

```
public class ProxyServiceManager {
  private HashMap<String, String> pluginServices = null;

  private static ProxyServiceManager instance = null;

  private ProxyServiceManager() {
    pluginServices = new HashMap<String, String>();
    pluginServices.put("jianqiang.com.plugin1.
    TestService1", "jianqiang.com.hostapp.
    ProxyService1");
    pluginServices.put("jianqiang.com.plugin2.
    TestService2", "jianqiang.com.hostapp.
    ProxyService1");
```

```
    pluginServices.put("jianqiang.com.plugin1.
    TestService3", "jianqiang.com.hostapp.
    ProxyService1");
  }

  public static ProxyServiceManager getInstance() {
    if(instance == null)
      instance = new ProxyServiceManager();

    return instance;
  }

  public String getProxyServiceName(String className) {
    return pluginServices.get(className);
  }
}
```

2) In the *MainActivity* of the HostApp, let's launch the *TestService1* of Plugin1:

```
public void startService1InPlugin1(View view) {
  try {
    Intent intent = new Intent();

    String serviceName = pluginItem1.packageInfo.
    packageName + ".TestService1";
    String proxyServiceName = ProxyServiceManager.
    getInstance().getProxyServiceName(serviceName);
    intent.setClass(this, Class.forName
    (proxyServiceName));

    intent.putExtra(AppConstants.EXTRA_DEX_PATH,
    pluginItem1.pluginPath);
    intent.putExtra(AppConstants.EXTRA_CLASS,
    serviceName);

    startService(intent);

  } catch (ClassNotFoundException e) {
    e.printStackTrace();
  }
}
```

Now, we create the 1:1 mapping between the *Service* of the HostApp and *ProxyService* of the plug-in.

9.2.4 The Last Solution for *Service* Plug-Ins: Integration with *Dynamic-Proxy* and *Static-Proxy**

Can I use a *StubService* to handle multiple *Services* in a plug-in? I don't want to pre-declare a lot of *StubServices* as placeholders.

It's time to use the hook technology introduced in Chapter 8.

9.2.4.1 Parse Service *in the Plug-In*

First of all, we still need to use the class *BaseDexClassLoaderHookHelper* to merge all the *dex* files of the HostApp and plug-in together, we have already introduced these techniques in Section 8.2.3.

Second, the plug-in is also an *apk* file and has an *AndroidManifest.xml*, which defines the information of all app components, like *Activity* and *Service*. We can get this information from the plug-in when the HostApp starts, shown as follows:

```
public void preLoadServices(File apkFile) throws
Exception {
    Object packageParser = RefInvoke.createObject
    ("android.content.pm.PackageParser");

    // First call parsePackage to get the
    corresponding Package object to the Apk object
    Object packageObj = RefInvoke.invokeInstanceMetho
    d(packageParser, "parsePackage",
        new Class[] {File.class, int.class},
        new Object[] {apkFile, PackageManager.
        GET_SERVICES});

    // Read the services field in the Package object
    // The next thing to do is to obtain ServiceInfo
    corresponding to Service from List<Service>
    List services = (List) RefInvoke.getField
    Object(packageObj, "services");
```

---

* For the example code of this section, please refer to https://github.com/Baobaojianqiang/ ServiceHook3 and for the example code of this section, please refer to https://github.com/ Baobaojianqiang/ServiceHook4

```
// Invoke function generateServiceInfo to convert
PackageParser.Service to ServiceInfo
Class<?> packageParser$ServiceClass = Class.
forName("android.content.pm.PackageParser$Service");
Class<?> packageUserStateClass = Class.
forName("android.content.pm.PackageUserState");

int userId = (Integer) RefInvoke.
invokeStaticMethod("android.os.UserHandle",
"getCallingUserId");
Object defaultUserState = RefInvoke.createObject
("android.content.pm.PackageUserState");


// Parse the intent corresponding Service component
for (Object service : services) {
  // need to invoke android.content.pm.PackagePar
  ser#generateActivityInfo(android.content.
  pm.ActivityInfo, int, android.content.
  pm.PackageUserState, int)
  ServiceInfo info = (ServiceInfo) RefInvoke.invo
  keInstanceMethod(packageParser,
  "generateServiceInfo",
      new Class[] {packageParser$ServiceClass,
      int.class, packageUserStateClass, int.
      class},
      new Object[] {service, 0, defaultUserState,
      userId});

  mServiceInfoMap.put(new ComponentName(info.
  packageName, info.name), info);
 }
}
```

The method *parsePackage()* of *PackageParser* obtains the *Service* collection from the plug-in according to the path of the plug-in *apk*. But in this collection, the type of each *Service* is *android.content. pm.PackageParser$Service*, which is not visible to app developers. So, we use reflection syntax to convert each element in this collection to an instance of *ServiceInfo*. *ServiceInfo* is visible to app developers. The convert method is *generateServiceInfo()* of *PackageParser*; it returns a new collection storing the instances of the *Services* from the plug-in.

9.2.4.2 Create a Service Object Using Reflection

In the source code of the Android system, a *Service* object is created using the method *handleCreateService()* of *ActivityThread*, shown as follows:

```
private void handleCreateService(CreateServiceData
data) {
  LoadedApk packageInfo = getPackageInfoNoCheck(
    data.info.applicationInfo, data.compatInfo);
  Service service = null;
  java.lang.ClassLoader cl = packageInfo.
  getClassLoader();
  service = (Service) cl.loadClass(data.info.name).
  newInstance();

  service.onCreate();
  mServices.put(data.token, service);
}
```

In the method *handleCreateService()*, a *Service* object is created using reflection, executes its method *onCreate()*, and stores this object in a collection, *mServices*. All the *Services* of the app are stored in this collection.

In Section 8.3.3.1, we introduced how to read the *Service* collection of a plug-in. Each element of this collection is a *ServiceInfo* object.

Now let's try to convert this *ServiceInfo* object to a *Service* object.

The general idea is to create a parameter *data* using reflection; the type of *data* is *CreateServiceData*. Then we can invoke the method *handleCrea teService(CreateServiceData data)* of *ActivityThread*.

As mentioned earlier, there is a collection, *mServices*, in the *ActivityThread*, which stores all the *Services* of the current app. When we invoke the method *handleCreateService()* of *ActivityThread*, we actually add a new *Service* in the collection *mServices*.

Now we are using the "That" framework. All the components of the plug-in are puppets; these "puppets" don't have life, so the *Services* of the plug-in shouldn't exist in the collection *mServices* of the *ActivityThread*. We need to remove it from the collection *mServices* and save these *Services* in a new collection, *mServiceMap*, defined by ourselves.

Let's have a look at the code implementation:

```
private void proxyCreateService(ServiceInfo
serviceInfo) throws Exception {
  IBinder token = new Binder();
```

```
// Create CreateServiceData object, used as a
parameter passed to the ActivityThread
handleCreateService
Object createServiceData = RefInvoke.
createObject("android.app.ActivityThread$CreateSer
viceData");

RefInvoke.setFieldObject(createServiceData,
"token", token);

// Write info object
// This change is for that, LoadedApk will be the
main program of the ClassLoader when load the
Class, we choose Hook BaseDexClassLoader way to
load the Plug-In
serviceInfo.applicationInfo.packageName =
UPFApplication.getContext().getPackageName();
RefInvoke.setFieldObject(createServiceData,
"info", serviceInfo);

// Get the default compatibility configuration
Object defaultCompatibility = RefInvoke.
getStaticFieldObject("android.content.res.
CompatibilityInfo", "DEFAULT_COMPATIBILITY_INFO");
// Write compatInfo field
RefInvoke.setFieldObject(createServiceData,
"compatInfo", defaultCompatibility);


// private void handleCreateService(CreateService
Data data) {
Object currentActivityThread = RefInvoke.
getStaticFieldObject("android.app.ActivityThread",
"sCurrentActivityThread");
RefInvoke.invokeInstanceMethod(currentActivity
Thread, "handleCreateService",
    createServiceData.getClass(),
    createServiceData);

// The Service object created by
handleCreateService has no return value but is
stored in the mServices field of ActivityThread.
Here we manually get it
```

```
Map mServices = (Map) RefInvoke.getFieldObject(cur
rentActivityThread, "mServices");
Service service = (Service) mServices.get(token);

// After getting it, remove the service, we just
borrow the flowers
mServices.remove(token);

// Store this Service
mServiceMap.put(serviceInfo.name, service);
}
```

9.2.4.3 ProxyService and ServiceManager

Now we have our own *mServiceMap* of *ServiceManager* to store all the *Services* of the plug-in, we can implement the plug-in solution of *Service*.

Figure 9.4 shows the flowchart of *startService()* in the plug-in solution.

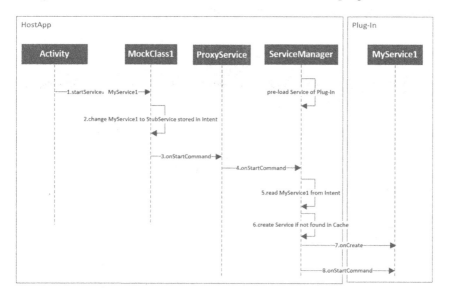

FIGURE 9.4　Flowchart of *startService()*.

First, let's focus on step 2, *MockClass1* is responsible for intercepting the method *startService()* and *stopService()*, and hooking *AMN* to replace the original *MyService1* with *ProxyService*. Refer to Chapter 6 for this technique.

ProxyService is started in this way, but it cannot be used for multiple *Services* in the plug-in at the same time, so it is necessary to write a *Singleton ServiceManager,* which is responsible for managing multiple *Services* of the plug-in.

The lifecycle method *onStartCommand()* of *ProxyService* will invoke the method *onStartCommand()* of *ServiceManager. ServiceManager* will analyses the *intent* in the parameter and pick up *MyService1* from the *intent.* Then *ServiceManager* will check if *MyService1* exists in *mService-Map.* If not found (meaning this is the first time to start *MyService1*), it will use the method *proxyCreateService()* to create an instance of *MyService1*

The instance of *MyService1* created in this way is only a normal class, and it does not have lifecycle methods like *onCreate(), onStartCommand(), onDestroy(),* and so on. Not all these lifecycle methods can be invoked by the Android system, so we have to call methods like *onStartCommand()* of *MyService1* manually.

That's the whole process of the method *startService().* Next, let's have a look at the process of the method *stopService().* Figure 9.5 shows the flow-chart of *stopService()* in the plug-in solution.

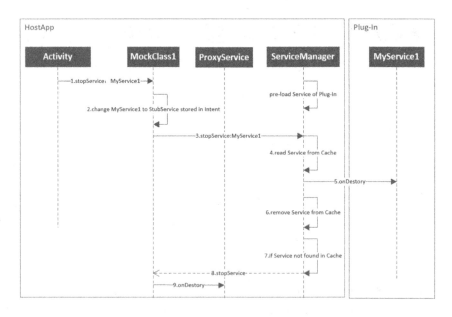

FIGURE 9.5 Flowchart of *stopService().*

In the plug-in solution for the method *stopService(),* we still use *MockClass1* to intercept the method *stopService*() of *AMN.*

We can not replace *MyService1* of the plug-in with *ProxyService* of the HostApp directly; we should invoke the method *stopService()* of

ServiceManager to invoke the method *onDestroy()* of *MyService1*, and then remove *MyService1* from *mServiceMap*.

Finally, we have to check if *mServiceMap* is empty. If empty, we should destroy *ProxyService*. But we can't invoke the method *onDestroy()* of *ProxyService* directly; we should invoke the method *stopService()* of *Context* to notify the Android system to destroy *ProxyService*.

Now, we have introduced the plug-in solution for *startService()* and *stopService()*. We use only one *ProxyService* to correspond to multiple *Services* of the plug-in. The code is shown as follows:

1) *MockClass1*

```java
class MockClass1 implements InvocationHandler {

  private static final String TAG = "MockClass1";

  // Alias ProxyService package name
  private static final String stubPackage =
  "jianqiang.com.activityhook1";

  Object mBase;

  public MockClass1(Object base) {
    mBase = base;
  }

  @Override
  public Object invoke(Object proxy, Method method,
  Object[] args) throws Throwable {

    Log.e("bao", method.getName());

    if ("startService".equals(method.getName())) {
      // Only intercept this function
      // Replace parameters and do whatever you want;
    // even replace the original ProxyService to start
    another Service.

      // Find the first Intent object in the parameter
      int index = 0;
      for (int i = 0; i < args.length; i++) {
        if (args[i] instanceof Intent) {
```

```
      index = i;
      break;
    }
  }
}

//get ProxyService form UPFApplication.
pluginServices
Intent rawIntent = (Intent) args[index];

// The package name of the proxy service, which
is our own package name
String stubPackage = UPFApplication.getContext().
getPackageName();


// replace Plug-In Service of ProxyService
ComponentName componentName = new
ComponentName(stubPackage, ProxyService.class.
getName());
Intent newIntent = new Intent();
newIntent.setComponent(componentName);

// The TargetService we originally started to
save first
newIntent.putExtra(AMSHookHelper.EXTRA_TARGET_
INTENT, rawIntent);

// Replace Intent, cheat AMS
args[index] = newIntent;

Log.d(TAG, "hook success");
return method.invoke(mBase, args);
} else if ("stopService".equals(method.getName())) {
// Only intercept this function
// Replace parameters and do whatever you want;
// even replace the original ProxyService to start
another Service.
// Find the first Intent object in the parameter
int index = 0;
for (int i = 0; i < args.length; i++) {
  if (args[i] instanceof Intent) {
    index = i;
```

```
      break;
    }
  }

  Intent rawIntent = (Intent) args[index];
  Log.d(TAG, "hook success");
  return ServiceManager.getInstance().
  stopService(rawIntent);
}

  return method.invoke(mBase, args);
}
}
```

2) *ProxyService*

```
public class ProxyService extends Service {

  private static final String TAG = "ProxyService";

  @Override
  public void onCreate() {
    Log.d(TAG, "onCreate() called");
    super.onCreate();
  }

  @Override
  public int onStartCommand(Intent intent, int flags,
int startId) {
    Log.d(TAG, "onStart() called with " + "intent = ["
+ intent + "], startId = [" + startId + "]");

    // 分发Service
    ServiceManager.getInstance().onStartCommand
    (intent, flags, startId);
    return super.onStartCommand(intent, flags, startId);
  }

  @Override
  public IBinder onBind(Intent intent) {
    return null;
  }
```

```
  @Override
  public void onDestroy() {
    Log.d(TAG, "onDestroy() called");
    super.onDestroy();
  }
}
```

3) *ServiceManager* (the method *proxyCreateService()* is in Section 9.2.4.2 and the method *preLoadServices()* is in Section 9.2.4.1)

```
public final class ServiceManager {

  private static final String TAG = "ServiceManager";

  private static volatile ServiceManager sInstance;

  private Map<String, Service> mServiceMap = new
  HashMap<String, Service>();

  // Storage Plug-In Service information
  private Map<ComponentName, ServiceInfo>
  mServiceInfoMap = new HashMap<ComponentName,
  ServiceInfo>();

  public synchronized static ServiceManager
  getInstance() {
    if (sInstance == null) {
      sInstance = new ServiceManager();
    }
    return sInstance;
  }

  /**
   * Start a plug-in Service; If the Service has not been
   started yet, a new plug-in Service will be created
   * @param proxyIntent
   * @param startId
   */
  public int onStartCommand(Intent proxyIntent, int
  flags, int startId) {

    Intent targetIntent = proxyIntent.getParcelableExt
    ra(AMSHookHelper.EXTRA_TARGET_INTENT);
```

```
    ServiceInfo = selectPluginService(targetIntent);

    try {
      if (!mServiceMap.containsKey(serviceInfo.name)) {
        // Service does not exist yet, first create it
        proxyCreateService(serviceInfo);
      }

      Service service = mServiceMap.get(serviceInfo.
      name);
      return service.onStartCommand(targetIntent,
      flags, startId);
    } catch (Exception e) {
      e.printStackTrace();
      return -1;
    }
}

/**
 * Stop a Plug-In Service, ProxyService will stop
 when all Plug-In services are stopped
 * @param targetIntent
 * @return
 */
public int stopService(Intent targetIntent) {
  ServiceInfo serviceInfo = selectPluginService(targ
  etIntent);
  if (serviceInfo == null) {
    Log.w(TAG, "cannot found service: " + targetIntent.
    getComponent());
    return 0;
  }
  Service service = mServiceMap.get(serviceInfo.name);
  if (service == null) {
    Log.w(TAG, "cannot run, stopped multiple times");
    return 0;
  }

  service.onDestroy();

  mServiceMap.remove(serviceInfo.name);
  if (mServiceMap.isEmpty()) {
```

```
      // Without Service, this mServiceMap does not
      need to exist
      Log.d(TAG, "service all stopped, stop proxy");
      Context appContext = UPFApplication.getContext();
      appContext.stopService(new Intent().
      setComponent(new ComponentName(appContext.
      getPackageName(), ProxyService.class.
      getName())));
    }
    return 1;
  }


  /**
   * Select the matching ServiceInfo
   * @param pluginIntent Plug-In Intent
   * @return
   */
  private ServiceInfo selectPluginService(Intent
  pluginIntent) {
    for (ComponentName componentName : mService
    InfoMap.keySet()) {
      if (componentName.equals(pluginIntent.
      getComponent())) {
        return mServiceInfoMap.get(componentName);
      }
    }
    return null;
  }
}
```

9.2.4.4 bindService() *and* unbindService()*

We have learned that the plug-in solution of *startService()* and *stopService()*, and the solution of *bindService()* and *unbindService()*, is relatively simple.

Figures 9.6 and 9.7 shows the flowchart of *bindService()* and *unbindService()* in the plug-in solution.

The plug-in solution of *bindService()* is the same as *startService()*, and the plug-in solution of *unbindService()* is the same as *stopService()*.

---

* For the example code of this section, please refer to https://github.com/Baobaojianqiang/ServiceHook4

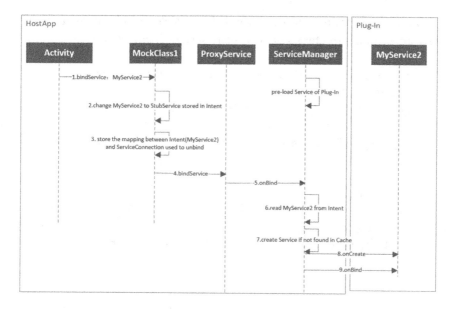

FIGURE 9.6 Flowchart of *bindService()*.

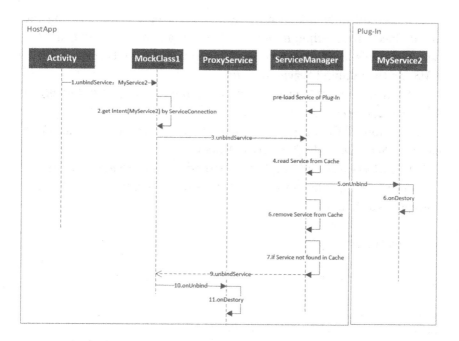

FIGURE 9.7 Flowchart of *unbindService()*.

The differences are that the methods *bindService()* and *unbindService()* have an extra parameter *conn*; its type is *ServiceConnection*.

```
ServiceConnection conn = new ServiceConnection() {
    @Override
    public void onServiceConnected(ComponentName
    name, IBinder service) {
      Log.d("baobao", "onServiceConnected");
    }

    @Override
    public void onServiceDisconnected(ComponentName
    componentName) {
      Log.d("baobao", "onServiceDisconnected");
    }
};
```

```
bindService(intent, conn, Service.BIND_AUTO_CREATE);
unbindService(conn);
```

The method *unbindService()* doesn't have a parameter carrying the *intent* object. The method *unbindService()* has a parameter *conn*, and the type of *conn* is *ServiceConnection*. The app process sends *conn* to *AMS*. *AMS* can then unbind *Service* by *conn*.

But when *ServiceManager* invokes the method *onUnbind(intent)* of *MyService2*, it needs the *intent* parameter.

So, we store all the 1:1 mappings between *conn* and *intent* in a *HashMap* named *mServiceInfoMap2* in *ServiceManager*.

When *MockClass1* intercepts the method *unbindService()*, we can get the corresponding *intent* by *conn*, and pass this *intent* to *ServiceManager*. The code is as follows:

1) *MockClass1*

```
package jianqiang.com.activityhook1.ams_hook;

import android.content.ComponentName;
import android.content.Intent;
import android.util.Log;

import java.lang.reflect.InvocationHandler;
```

```java
import java.lang.reflect.Method;

import jianqiang.com.activityhook1.ProxyService;
import jianqiang.com.activityhook1.ServiceManager;
import jianqiang.com.activityhook1.UPFApplication;

class MockClass1 implements InvocationHandler {

  private static final String TAG = "MockClass1";

  Object mBase;

  public MockClass1(Object base) {
    mBase = base;
  }

  @Override
  public Object invoke(Object proxy, Method method,
  Object[] args) throws Throwable {

    Log.e("bao", method.getName());
    if("bindService".equals(method.getName())) {
      // Only intercept this method
      // Replace parameters and do whatever you want;
    // replace the original ProxyService to start
    another Service.
      // Find the first Intent object in the parameter
      int index = 0;
      for (int i = 0; i < args.length; i++) {
        if (args[i] instanceof Intent) {
          index = i;
          break;
        }
      }
    }

    //get ProxyService form UPFApplication.
    pluginServices
    Intent rawIntent = (Intent) args[index];

    //store intent-conn
    ServiceManager.getInstance().mServiceMap2.
    put(args[4], rawIntent);
```

```
    // The package name of the proxy service, which
    is our own package name
    String stubPackage = UPFApplication.getContext().
    getPackageName();

    // replace Plug-In Service of ProxyService
    ComponentName componentName = new ComponentName
    (stubPackage, ProxyService.class.getName());
    Intent newIntent = new Intent();
    newIntent.setComponent(componentName);

    // The TargetService we originally started to
    save first
    newIntent.putExtra(AMSHookHelper.EXTRA_TARGET_
    INTENT, rawIntent);

    // Replace Intent, cheat AMS
    args[index] = newIntent;

    Log.d(TAG, "hook success");
    return method.invoke(mBase, args);
  } else if("unbindService".equals(method.getName())) {
    Intent rawIntent = ServiceManager.getInstance().
    mServiceMap2.get(args[0]);
    ServiceManager.getInstance().onUnbind(rawIntent);
    return method.invoke(mBase, args);
  }

  return method.invoke(mBase, args);
  }
}
```

2) *ProxyService*

```
public class ProxyService extends Service {

  private static final String TAG = "ProxyService";

  @Override
  public void onCreate() {
    Log.d(TAG, "onCreate() called");
    super.onCreate();
  }
```

```
@Override
public IBinder onBind(Intent intent) {
  Log.e("jianqiang", "Service is binded");

  return ServiceManager.getInstance().
  onBind(intent);
}

@Override
public boolean onUnbind(Intent intent) {
  Log.e("jianqiang", "Service is unbinded");

  return super.onUnbind(intent);
}
}
```

3) *ServiceManager*

```
public final class ServiceManager {

  private Map<String, Service> mServiceMap = new
  HashMap<String, Service>();

  //store intent-conn
  public Map<Object, Intent> mServiceMap2 = new
  HashMap<Object, Intent>();

  public IBinder onBind(Intent proxyIntent) {

    Intent targetIntent = proxyIntent.getParcelableExt
    ra(AMSHookHelper.EXTRA_TARGET_INTENT);
    ServiceInfo serviceInfo = selectPluginService
    (targetIntent);

    try {
      if (!mServiceMap.containsKey(serviceInfo.name)) {
        // Service does not exist yet, first create
        proxyCreateService(serviceInfo);
      }

      Service service = mServiceMap.get(serviceInfo.name);
      return service.onBind(targetIntent);
    } catch (Exception e) {
```

```
      e.printStackTrace();
      return null;
    }
  }


  /**
   * Stop a Plug-In Service, ProxyService will stop
  when all Plug-In services are stopped
   * @param targetIntent
   * @return
   */
  public boolean onUnbind(Intent targetIntent) {
    ServiceInfo serviceInfo = selectPluginService
    (targetIntent);
    if (serviceInfo == null) {
      Log.w(TAG, "cannot found service: " + targetIntent.
      getComponent());
      return false;
    }
    Service service = mServiceMap.get(serviceInfo.name);
    if (service == null) {
      Log.w(TAG, "cannot run, stopped multiple times");
      return false;
    }

    service.onUnbind(targetIntent);

    mServiceMap.remove(serviceInfo.name);
    if (mServiceMap.isEmpty()) {
      // Without Service, this mServiceMap does not
      need to exist
      Log.d(TAG, "service all stopped, stop proxy");
      Context appContext = UPFApplication.getContext();
      appContext.stopService(
          new Intent().setComponent(new ComponentName
          (appContext.getPackageName(), ProxyService.
          class.getName())));
    }
    return true;
  }
}
```

Up until now, the "That" framework supports *Service*.

9.2.5 *Static-Proxy* in *BroadcastReceiver**

How to use *Static-Proxy* in *BroadcastReceiver*?

BroadcastReceiver is a normal class; it has only one method *onReceive()*.

Let's create a *ProxyReceiver* in the HostApp; it's a *Receiver*. When we invoke the method *sendBroadcast(Intent intent)* to fire the method *onReceive()* of *ProxyReceiver*, the parameter *intent* carries the information about the *Receiver* of Plug-In. When *ProxyReceiver* receives this information, it will invoke the method *sendBroadcast* again, to fire the method *onReceive()* of *ProxyReceiver*. Let's have a look at how to implement this idea.

Because both *Activity* and *Service* inherit from *Context*, we can invoke the method *getClassLoader()* and *getDir()* of *Context* in *Activity* and *Service*. We can pass *dexPath* of the plug-in as a parameter to *ProxyActivity* and *ProxyService*, and then create a *ClassLoader* for the plug-in to invoke the method *loadClass()* of *ClassLoader* to get the corresponding *Activity* and *Service*.

But we find it doesn't work for *Receiver*, because *Receiver* does not have a *Context* inside; we can't use the syntax like *getClassLoader()* or *getDir()*, so it's no use to pass *dexPath* to *ProxyReceiver*.

We can also generate all the *ClassLoaders* of the plug-ins at one time and put them into a *HashMap*. When *ProxyReceiver* wants to use the *ClassLoader* of the PLUG-IN, we can fetch it from this *HashMap* using the name of the plug-in.

1) Declare *ProxyReceiver* in *AndroidManifest.xml*:

```
<receiver android:name=".ProxyReceiver">
  <intent-filter>
    <action android:name="baobao2" />
  </intent-filter>
</receiver>
```

2) Prepare *MyClassLoaders* which carries all the plug-ins:

```
public class MyClassLoaders {
  public static final HashMap<String, DexClassLoader>
  classLoaders = new HashMap<String, DexClassLoader>();
}
```

---

* For the example code of this section, please refer to https://github.com/Baobaojianqiang/That3.4

302 ■ Android App-Hook and Plug-In Technology

In the *MainActivity* of the HostApp, pre-load all the *ClassLoaders* of plug-ins into *MyClassLoaders*:

```
pluginItem1 = generatePluginItem("plugin1.apk");
pluginItem2 = generatePluginItem("plugin2.apk");

private PluginItem generatePluginItem(String
apkName) {
  File file = getFileStreamPath(apkName);
  PluginItem item = new PluginItem();
  item.pluginPath = file.getAbsolutePath();
  item.packageInfo = DLUtils.getPackageInfo(this,
  item.pluginPath);

  String mDexPath = item.pluginPath;

  File dexOutputDir = this.getDir("dex", Context.
  MODE_PRIVATE);
  final String dexOutputPath = dexOutputDir.
  getAbsolutePath();
  DexClassLoader dexClassLoader = new DexClassLoader
  (mDexPath,
      dexOutputPath, null, getClassLoader());

  MyClassLoaders.classLoaders.put(apkName, dexClass
  Loader);

  return item;
}
```

Now we can use any *ClassLoader* directly in *ProxyReceiver*:

```
public class ProxyReceiver extends BroadcastReceiver {

  private static final String TAG = "ProxyService";

  private String mClass;
  private String pluginName;
  private IRemoteReceiver mRemoteReceiver;

  @Override
  public void onReceive(Context context, Intent intent) {
```

```
Log.d(TAG, TAG + " onReceive");

pluginName = intent.getStringExtra(AppConstants.
EXTRA_PLUGIN_NAME);
mClass = intent.getStringExtra(AppConstants.
EXTRA_CLASS);

try {
    //Get Plug-In receiver object by reflection
    Class<?> localClass = MyClassLoaders.
    classLoaders.get(pluginName).
    loadClass(mClass);
    Constructor<?> localConstructor = localClass.
    getConstructor(new Class[] {});
    Object instance = localConstructor.
    newInstance(new Object[] {});

    mRemoteReceiver = (IRemoteReceiver) instance;
    mRemoteReceiver.setProxy(this);
    mRemoteReceiver.onReceive(context, intent);

} catch (Exception e) {
    e.printStackTrace();
}
}
}
```

3) Send a broadcast to the *ProxyReceiver*, pass the package name of the *apk* and the name of the *Receiver* from the plug-in to the *ProxyReceiver*. Now we needn't pass *dexPath* to the *ProxyReceiver* anymore.

```
public void notifyReceiver1(View view) {
  Intent intent = new Intent(MainActivity.ACTION);
  intent.putExtra(AppConstants.EXTRA_PLUGIN_NAME,
  "plugin1.apk");
  intent.putExtra(AppConstants.EXTRA_CLASS,
  "jianqiang.com.plugin1.TestReceiver1");
  sendBroadcast(intent);
}

public void notifyReceiver2(View view) {
  Intent intent = new Intent(MainActivity.ACTION);
```

```
    intent.putExtra(AppConstants.EXTRA_PLUGIN_NAME,
    "plugin2.apk");
    intent.putExtra(AppConstants.EXTRA_CLASS,
    "jianqiang.com.plugin2.TestReceiver2");
    sendBroadcast(intent);
}
```

4) All the *Receivers* of the plug-in must inherit the parent class *BasePluginReceiver* and implement the interface *IRemoteReceiver*.

```
public interface IRemoteReceiver {
  public void onReceive(Context context, Intent intent);

  public void setProxy(BroadcastReceiver proxyReceiver);
}

public class BasePluginReceiver extends Broadcast
Receiver implements IRemoteReceiver{

  public static final String TAG = "BasePluginReceiver";
  private BroadcastReceiver that;

  @Override
  public void setProxy(BroadcastReceiver proxyReceiver) {
    that = proxyReceiver;
  }

  @Override
  public void onReceive(Context context, Intent intent) {

  }
}

public class TestReceiver1 extends BasePluginReceiver {

  private static final String TAG = "TestReceiver1";

  @Override
  public void onReceive(Context context, Intent intent) {
    Log.e(TAG, "TestReceiver1 onReceive");
  }
}
```

Up until now, the plug-in solution for *BroadcastReceiver* is completed, and there is a one-to-many relationship between the *ProxyReceiver* and the *Receivers* of the plug-in.

Unfortunately, the "That" framework only supports *Dynamic Receivers*. It doesn't support *Static Receivers*.

9.3 SUMMARY

We spent a lot of time to introduce the "That" framework. Although "That" is no longer maintained by its creator, it's still being used by many companies in their enterprise-level apps.

"Puppet" is the most vivid description of this plug-in framework.

CHAPTER 10

Related Plug-In Techniques

U P UNTIL NOW, WE have introduced the plug-in solutions of four components. But it's not enough; we still have a lot of problems to face. In this chapter, we will introduce how to resolve these problems in detail.

10.1 RESOLVE THE CONFLICTS BETWEEN RESOURCES OF THE PLUG-INS

We have already introduced how to load *Resources* from the plug-in in Chapter 7.

Each resource has an ID in *R.java*, such as 0x7f00010002. Because the HostApp and plug-in are packaged separately, the ID value in the HostApp may be the same as the ID value of plug-in. In this scenario, the app can't load the correct resource

We'll resolve this problem of ID conflict in this section. Let us focus on the command *aapt* during the app packaging process.

10.1.1 The Process of App Packaging

Before 2014, the packaging of an Android app was based on Ant, and we needed to know each step of the packaging process, including which command was executed, and which parameters were required for this command.

Later, Gradle was widely used in Android packaging; the packaging process was simplified to a simple configuration file in Gradle. Some people are familiar with Gradle, but they don't know the commands like *aapt* or *zipalign*.

Figure 10.1 shows the whole process of Android app packaging. Let's introduce these steps one by one:

1) *aapt*. Generate two files, *R.java* corresponds to the resources in the folder *Res*, *Manifest.java* corresponds to the *AndroidManifest. xml*.

2) *AIDL*. Generate a Java file for each *AIDL* file written by the app developers into the app.

3) *javac*. Compile the Java code to generate a lot of *.class* files.

4) *Proguard*. Obfuse code and generate a *mapping.txt*. This step is optional.

5) *dex*. Convert all the *.class* files into *dex* files.

6) *aapt*. Package the resources in the folder *Res* and the files in the folder *Assets* into a *zip* file with the suffix *.ap_*. This is another important feature of the command *aapt*.

7) *apkbuilder*. Package the *dex* files, the files with the suffix *.ap_*, and the *AndroidManifest.xml* into an *apk* file, which is unsigned.

8) *jarsigner*. Signature this unsigned *apk* file.

9) *zipalign*. Execute this command to reduce the usage of the memory during runtime.

In these nine steps, we focus on the first step, the generation of the *R.java*. We can control the generation process of the resource ID.

10.1.2 Hook aapt*

10.1.2.1 Modify and Generate a New aapt Command

As we know, the Android system will generate the resource ID in *R.java* for each resource in the folder *Res*.

---

* For the example code from this section, please refer to https://github.com/Baobaojianqiang/AAPT

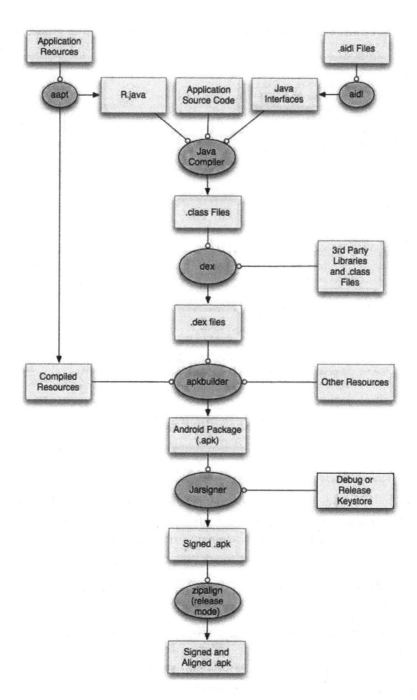

FIGURE 10.1   Android app packaging process.

Each resource in the folder *Res* has a *HEX* value in *R.java*, shown as follows:

```
public final class R {
    public static final class anim {
        public static final int abc_fade_in=0x7f050000;
        public static final int abc_fade_out=0x7f050001;
        public static final int abc_grow_fade_in_from_bottom=0x7f050002;
        public static final int abc_popup_enter=0x7f050003;
        public static final int abc_popup_exit=0x7f050004;
        public static final int abc_shrink_fade_out_from_bottom=0x7f050005;
        public static final int abc_slide_in_bottom=0x7f050006;
        public static final int abc_slide_in_top=0x7f050007;
        public static final int abc_slide_out_bottom=0x7f050008;
        public static final int abc_slide_out_top=0x7f050009;
    }

public static final class id {
    public static final int action0=0x7f0b006d;
    public static final int action_bar=0x7f0b0047;
    public static final int action_bar_activity_content=0x7f0b0000;
    public static final int action_bar_container=0x7f0b0046;
    public static final int action_bar_root=0x7f0b0042;
    public static final int action_bar_spinner=0x7f0b0001;
    public static final int action_bar_subtitle=0x7f0b0025;
    public static final int action_bar_title=0x7f0b0024;
    public static final int action_container=0x7f0b006a;
    public static final int action_context_bar=0x7f0b0048;
    public static final int action_divider=0x7f0b0071;
    public static final int action_image=0x7f0b006b;
```

The hex value such as 0x7f0b006d consists of three parts: *PackageId* + *TypeId* + *EntryId*:

- *PackageId*: This field is always 0x7f.

- *TypeId*: Stands for the resource type. Figure 10.2 lists all the types of resources; we are familiar with *layout, string, drawable,* and so on. The value of *TypeId* increments from 1; for example, attr=0x01, drawable=0x02.

- *EntryId*: The ID value of the resource under this type, incremented from 0.

For example, 0x7f0b006d, *PackageId* is 0x7f, *TypeId* is 0b00, and *EntryId* is 6d.

In the plug-in project, we find the resource ID of the HostApp is always the same as the resource ID of the plug-in.

To resolve this problem, we can set different *PackageId* for different plug-ins. For example, the HostApp has two plug-ins, Plugin1 and Plugin2. We set the *PackageId* of Plugin1 to 0x40, set the *PackageId* of Plugin12 to

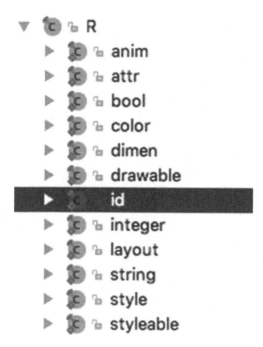

FIGURE 10.2   Structure of *R.java*.

0x40. The *PackageId* of the HostApp is always 0x7f, so this conflict won't occur anymore.

In the whole process of packaging, *aapt* is used to generate the resource ID for the resources in the folder *Res*. The default value of *PackageId* is 0x7f. We need to modify the source code of the *aapt* command.

The source code of *aapt* is located in the Android SDK, so search "0x7f" in the source code; we find the logic of 0x7f, shown as follows:

```
ResourceTable::ResourceTable(Bundle* bundle, const
String16& assetsPackage, ResourceTable::PackageType
type)
  : mAssetsPackage(assetsPackage)
  , mPackageType(type)
  , mTypeIdOffset(0)
  , mNumLocal(0)
  , mBundle(bundle)
{
  ssize_t packageId = -1;
  switch (mPackageType) {
    case App:
```

```
case AppFeature:
  packageId = 0x7f;
  break;

case System:
  packageId = 0x01;
  break;

case SharedLibrary:
  packageId = 0x00;
  break;

default:
  assert(0);
  break;
}


//omit some code below
}
```

The code above is the constructor of *ResourceTable* with a *Bundle* parameter. In the switch…case… statement, we find the app defaults to 0x7f, and we also find 0x00 and 0x01 are used by the Android system itself. So, we can't use 0x00 and 0x01 as the prefix of the resource ID of the plug-in.

The process of *aapt* is shown in Figure 10.3.

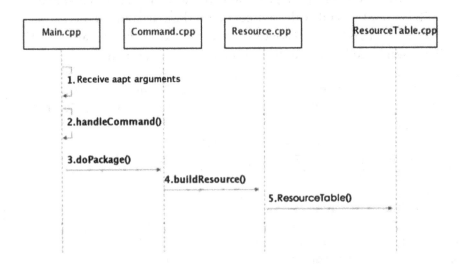

FIGURE 10.3   Flow of generating resource ID by *aapt*.

Next, let's talk about how to modify the code of *aapt* to modify the prefix of the resource ID of the plug-in, as below:

1) Add a new argument in the *aapt* command, pass the new value of the prefix such as 0x71 to the *aapt* command.

2) Pass 0x71 as a parameter to the constructor of *ResourceTable*.

3) In the constructor of *ResourceTable*, set 0x71 to the variable *PackageId*.

The implementation is as follows:

1) In the function *main* of *Main.cpp*, we add a new argument, *-PLUG-resource-id*:

```
else if(strcmp(cp, "-PLUG-resource-id") == 0){
  argc--;
  argv++;
  if (!argc) {
    fprintf(stderr, "ERROR: No argument supplied for
    '--PLUG-resource-id' option\n");
    wantUsage = true;
    goto bail;
  }
  bundle.setApkModule(argv[0]);
}
```

2) In *Bundle.h*, we add two methods: *getApkModule()* and *setApkModule()*:

```
//pass Plug-In prefix
const android::String8& getApkModule() const {return
mApkModule;}
void setApkModule(const char* str) { mApkModule=str;}
```

3) In the constructor of *ResourceTable*, append some code after the switch statement:

```
if(!bundle->getApkModule().isEmpty()){
  android::String8 apkmoduleVal=bundle->getApkModule();
  packageId=apkStringToInt(apkmoduleVal);
}
```

Now we compile and generate a new *aapt* command.

10.1.2.2 Using This New aapt *Command in the Project*

We replace the original *aapt* command with this new *aapt* command. But it's inconvenient for us to supply a new *appt* command for different versions of the Android platforms.

Another solution is to put this new *aapt* command into the plug-in project. We rename this command file as *aapt_mac* (it's compiled in mac) and place it in the root directory of the project, as shown in Figure 10.4.

FIGURE 10.4   Rename *aapt* tool to *aapt_mac*.

Then let's modify the Gradle file, shown as follows:

```
apply Plug-In: 'com.android.application'

import com.android.sdklib.BuildToolInfo
import java.lang.reflect.Method

Task modifyAaptPathTask = task('modifyAaptPath') << {
  android.applicationVariants.all { variant ->
    BuildToolInfo buildToolInfo = variant.
    androidBuilder.getTargetInfo().getBuildTools()
    Method addMethod = BuildToolInfo.class.
    getDeclaredMethod("add", BuildToolInfo.PathId.
    class, File.class)
```

```
      addMethod.setAccessible(true)
      addMethod.invoke(buildToolInfo, BuildToolInfo.
      PathId.AAPT, new File(rootDir, "aapt_mac"))
      println "[LOG] new aapt path = " + buildToolInfo.
      getPath(BuildToolInfo.PathId.AAPT)
  }
}

android {
  compileSdkVersion 25
  buildToolsVersion "25.0.3"

  defaultConfig {
    applicationId "jianqiang.com.testreflection"
    minSdkVersion 21
    targetSdkVersion 25
    versionCode 1
    versionName "1.0"
  }
  buildTypes {
    release {
      minifyEnabled false
      proguardFiles getDefaultProguardFile('proguard-
      android.txt'), 'proguard-rules.pro'
    }
  }

  preBuild.doFirst {
    modifyAaptPathTask.execute()
  }

  aaptOptions {
    aaptOptions.additionalParameters '--PLUG-
    resource-id', '0x71'
  }
}

dependencies {
  compile fileTree(dir: 'libs', include: ['*.jar'])
  testCompile 'junit:junit:4.12'
  compile 'com.android.support:appcompat-v7:25.2.0'
}
```

In the code above, we modify the path of *aapt* to the new path of *mac_aapt* in the root directory.

In addition, we set the prefix of the resources to 0x71. This means all the resource IDs in this *apk* will start with "0x71" after packaging.

10.1.3 public.xml*

Sometimes we need a custom control which is used in PluginA and PluginB. So, we write this control into the HostApp. PluginA and PluginB use this control and its resources in the HostApp

The resource ID of the HostApp is always changing as we add or remove resources to the HostApp. If the resource ID of the custom control changes, PluginA will not find this resource. So, we need to set the resource ID of this custom control to a fixed value.

When the version of Gradle is smaller than 1.3, we can define a file *public.xml* in the folder *Res/Values*, as follows:

```
<?xml version="1.0" encoding="utf-8" ?>
<resources>
  <public type = "string" name="string1" id =
  "0x7f050024"/>
</resources>
```

We define *string1* in *public.xml*, its value is 0x7f050024. When we use *R.string1* in the app, its value is always 0x7f050024.

We can also define a range from 0x7f02000f to 0x7f020001 for a special resource, as follows, don't ignore the space after "type" and "id":

```
<?xml version="1.0" encoding="utf-8" ?>
<resources>
    <public-padding name="my_" end="0x7f02000f"
    start="0x7f020001" type="drawable" />
</resources>
```

In Gradle 1.3 or the earlier versions, *public.xml* is not supported, even though we put this file in the folder *Res/Values* it doesn't work. But we can write some code into a Gradle file to implement this function.

Based on the project *ActivityHost1* introduced in the previous chapters, we need to modify three places, as follows:

---

* For the example code of this section, please refer to https://github.com/Baobaojianqiang/Apollo1.1

1) Modify *build.gradle* of *ActivityHook1*:

```
afterEvaluate {
  for (variant in android.applicationVariants) {
    def scope = variant.getVariantData().getScope()
    String mergeTaskName = scope.getMergeResourcesTask().
    name

    def mergeTask = tasks.getByName(mergeTaskName)

    mergeTask.doLast {
      copy {
        int i = 0
        println android.sourceSets.main.res.srcDirs
        from(android.sourceSets.main.res.srcDirs) {
          include 'values/public.xml'
          rename 'public.xml', (i++ == 0 ? "public.xml"
          : "public_${i}.xml")
        }

        into(mergeTask.outputDir)
      }
    }
  }
}
```

2) Open the file *string.xml* in the folder *Res/Values* of *ActivityHost1* and add one line:

```
<string name="string1">Test String</string>
```

3) Create *public.xml* in the folder *Res/Values*:

```
<?xml version="1.0" encoding="utf-8" ?>
<resources>
  <public type = "string" name="string1" id =
  "0x7f050024"/>
</resources>
```

Let's package the project *ActivityHost1*, and open *ActivityHost1.apk* by JadxGUI. We find that the value of *R.string.string1* is always 2131034148, which is a decimal value, corresponding to a *HEX* value of 0x7f050024 (Figure 10.5).

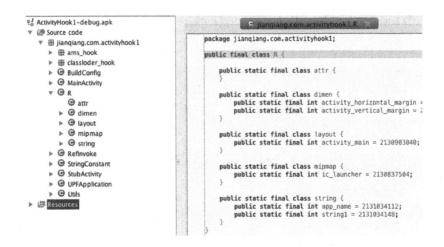

FIGURE 10.5   Resource ID in JadxGUI.

10.1.4 Plug-In Uses Resources in the HostApp*

The resource ID defined in *public.xml* of the HostApp is a fixed value, but how does the plug-in access the resource ID of the HostApp?

If we compile the HostApp as a *ClassLibrary*, we can add a reference to this *ClassLibrary* in the plug-in. So, plug-ins can visit any class or resource of the HostApp. We need to write a script in Gradle to compile the project's HostApp to a *jar* file and copy it into the project plug-in. We also need to change the keyword "compile" to "provide" in the Gradle file, to make sure the *jar* generated by the HostApp won't be compiled into the plug-in app.

We introduced the project Apollo1.1 in Section 10.1.3.

The project in this section is Apollo1.2; it's based on Apollo1.1. Let's begin our exploration.

1) Write a task in *build.gradle* of the HostApp to generate the *jar* file of *ActivityHost1*:

```
task buildJar(dependsOn: ["compileReleaseJavaWith
Javac"], type: Jar) {
  //final Jar name
  archiveName = "sdk2.jar"

  //the directory of needed resource to package
    def srcClassDir = [project.buildDir.absolutePath +
    "/intermediates/classes/release"]
```

---

* For the example code of this section, please refer to https://github.com/Baobaojianqiang/Apollo1.2

```
//directory of initial resource
  from srcClassDir
}
```

2) Execute *buildJar* (in the Gradle panel of Android Studio), to generate *sdk2.jar* under the folder */Build/Libs* of *ActivityHost1*.

3) Create a folder *sdk-jars* under the root directory of the TestActivity project and copy *sdk2.jar* into this folder

4) Add one line in *build.gradle* of the TestActivity (TestActivity is a plug-in project):

```
provided files('sdk-jars/sdk2.jar')
```

We use the keyword "provided" rather than "compile," meaning that the *sdk2.jar* file is only used at the time of compilation and the file *sdk2.jar* will not be compiled into the plug-in app.

After clicking "Sync" in the Gradle panel of Android Studio, we can use *StringConstant.string1* defined in *ActivityHost1* of the TestActivity project, shown as follows:

```
import jianqiang.com.activityhook1.StringConstant;

public class MainActivity extends Activity {

  @Override
  protected void onCreate(Bundle savedInstanceState) {
    super.onCreate(savedInstanceState);
    TextView tv = new TextVicw(this);
    tv.setText("baobao222");
    setContentView(tv);

    Log.d("baobao2", String.valueOf(StringConstant.
    string1));
  }
}
```

This section provides various solutions to resolve the conflicts of resource ID between the HostApp and plug-ins. Let's have a review:

Solution 1: Merge the resources of the HostApp and the plug-in together. We invoke the method *addAssetPath()* of *AssetManager* to merge these resources.

Solution 1 may cause conflicts of resource ID. There are several:

- Solution 1.1: Rewrite the *aapt* command, specify the prefix of the resource ID, such as 0x71, to ensure that the resource ID of the HostApp and plug-in will never conflict.

- Solution 1.2: After packaging the plug-in app, modify the resource ID in *R.java* and *resource.arsc*, for example, modify the default prefix 0x7f to 0x71, so that there will never be conflict between the HostApp and plug-in app.

- Solution 1.3: Specify the values of all the resource IDs in *public.xml*. But doing this is very troublesome. Every time we add a resource in the app, we need to maintain it in *public.xml*. So, this solution can only be used to fix several specific values.

Solution 2: If we don't merge resources in the process of packaging, we need to create an *AssetManager* for each plug-in. Each *AssetManager* calls the method *addAssetPath* by reflection to add the resources of the plug-in. When we navigate from the HostApp to plug-in, we should switch to the *AssetManager* of the plug-in. We introduced this solution in Chapter 5,

We focus on solution 1 in this section.

10.2 A PLUG-IN FRAMEWORK BASED ON *FRAGMENT*

In this section, I will introduce an ancient plug-in framework named *AndroidDynamicLoader*;[*] it's the first open source plug-in framework published by Yimin Tu of Dianping.com in July 2012. This framework loads *Fragment* of the plug-in dynamically.

Unlike the four components of the Android system, *Fragment* is only a simple class. All the four components need to interact with the *AMS*, while *Fragment* has no relationship with the *AMS*.

The *AndroidDynamicLoader* framework is based on Eclipse and Ant; it's a bit out of date. I use this framework in Android Studio and Gradle; it has a new name *Min18Fragment*, which is better for us to understand the spirit of this framework.

10.2.1 *AndroidDynamicLoader* Overview

The *AndroidDynamicLoader* framework was the first open source plug-in framework. It proposes a lot of ideas and put these ideas into practice.

---

[*] https://github.com/mmin18/AndroidDynamicLoader

1) Override the following four methods to load resources of the plug-in:

- *getAssets()*
- *getResources()*
- *getTheme()*

I introduced the "That" framework in Chapter 9; it also overrides these four methods.

2) Create each *ClassLoader* for the plug-in *apk*. When the HostApp loads the class of the plug-in, it will use the corresponding *ClassLoader*. I introduced this technique in Chapter 6.

10.2.2 A Simple Plug-In Sample Based on *Fragment**

Let's have a look at how to load *Fragment* of the Plug-In.

1) Declare *FragmentLoaderActivity* in the *AndroidManifest.xml* of the HostApp, and it is used as the container of *Fragment*, shown as follows:

```
<activity android:name=".FragmentLoaderActivity">
  <intent-filter>
  <action android:name="jianqiang.com.hostapp.VIEW" />
  <category android:name="android.intent.category.
  DEFAULT" />
  </intent-filter>
</activity>
```

2) All the navigation logic is in *FragmentLoaderActivity*. For example, we want to load *Fragment1*:

```
Intent intent = new Intent(AppConstants.ACTION);
intent.putExtra(AppConstants.EXTRA_DEX_
PATH,mPluginItems.get(position).pluginPath);
intent.putExtra(AppConstants.EXTRA_CLASS,
mPluginItems.get(position).packageInfo.packageName +
".Fragment1");
startActivity(intent);
```

---

* For the example code of this section, please refer to https://github.com/Baobaojianqiang/Min18Fragment

We need to pass two parameters to *FragmentLoaderActivity*. One is the path of the plug-in, the other is the full name of the *Fragment* to be loaded.

Since then, we can write all the page logic in this *FragmentLoaderActivity*, putting each page in the *Fragment* and using whatever we want.

3) According to the parameter *dexPath*, *Fragmentloaderactivity* can get the corresponding *ClassLoader* and *Resources*. *Fragmentloaderactivity* uses the corresponding *ClassLoader* to launch the *Fragment* of the plug-in.

```
//Reflects the Plugin's Fragment object
Class<?> localClass = dexClassLoader.
loadClass(mClass);
Constructor<?> localConstructor = localClass.
getConstructor(new Class[] {});
Object instance = localConstructor.newInstance
(new Object[] {});
Fragment f = (Fragment) instance;
FragmentManager fm = getFragmentManager();
FragmentTransaction ft = fm.beginTransaction();
ft.replace(R.id.container, f);
ft.commit();
```

Note that *R.id.container* refers to the layout *activity_fragment_loaderd*.

10.2.3 Jumping Between *Fragments**

We are more interested in how to jump between different *Fragments*. There are four scenarios:

From	To
Fragment in the HostApp	*Fragment* in Plugin1
Fragment in Plugin1	*Fragment* in Plugin1
Fragment in Plugin1	*Fragment* in Plugin2
Fragment in Plugin1	*Fragment* in the HostApp

---

* For the example code of this section, please refer to https://github.com/Baobaojianqiang/Min18Fragment2

The first scenario, jumping from the HostApp to Plugin1, we have already talked about in Section 10.2.2.

In this section we focus on the second scenario, jumping between different *Fragments* in an *Activity* of *Plugin1*. The following code implements the logic of jumping from *Fragment1* to *Fragment2* in *FragmentLoaderActivity*.

```
public class Fragment1 extends BaseFragment {
  @Override
  public View onCreateView(LayoutInflater inflater,
  ViewGroup container, Bundle savedInstanceState) {
    View view = inflater.inflate(R.layout.fragment1,
    container, false);
    view.findViewById(R.id.load_fragment2_btn).
    setOnClickListener(new View.OnClickListener() {
      @Override
      public void onClick(View arg0) {
        Fragment2 fragment2 = new Fragment2();
        Bundle args = new Bundle();
        args.putString("username", "baobao");
        fragment2.setArguments(args);
        getFragmentManager()
            .beginTransaction()
            .addToBackStack(null) //Add the current
            fragment to the back stack
            .replace(Fragment1.this.getContainerId(),
            fragment2).commit();
      }
    });
    return view;
  }
}
```

The corresponding layouts of the *FragmentLoaderActivity* are defined in the HostApp, but we want to use this layout in the plug-in, so we define the parent class *BaseFragment* in the MyPluginLibrary, which has a field *containerId*. All the *Fragments* inherit from *BaseFragment*. When we load *Fragment* in *FragmentLoaderActivity*, we need to pass *R.id.container* to the field *containerId* of *BaseFragment*:

```
BaseFragment f = (BaseFragment) instance;
f.setContainerId(R.id.container);
```

Now we can jump from one *Fragment* to another *Fragment* in the same *Activity* of the plug-in,

10.2.4 Jump from the Plug-In*

Let's talk about how to jump from the plug-in to the HostApp or other plug-ins.

First, when we jump from the HostApp to the plug-in, we should switch the *ClassLoader* of the HostApp to the *ClassLoader* of the plug-in. *AssetManager* and *Resources* have the same logic. When we jump from the HostApp to the plug-in, we should switch the *AssetManager* and *Resources* of the HostApp to the *AssetManager* and *Resources* of the plug-in. We can encapsulate this logic in *BaseActivity*, shown as follows:

```
public class BaseHostActivity extends Activity {
  private AssetManager mAssetManager;
  private Resources mResources;
  private Theme mTheme;

  protected String mDexPath;
  protected ClassLoader dexClassLoader;

  protected void loadClassLoader() {
    File dexOutputDir = this.getDir("dex", Context.
    MODE_PRIVATE);
    final String dexOutputPath = dexOutputDir.
    getAbsolutePath();
    dexClassLoader = new DexClassLoader(mDexPath,dexO
    utputPath, null, getClassLoader());
  }

  protected void loadResources() {
    try {
      AssetManager assetManager = AssetManager.class.
      newInstance();
      Method addAssetPath = assetManager.getClass().
      getMethod("addAssetPath", String.class);
      addAssetPath.invoke(assetManager, mDexPath);
      mAssetManager = assetManager;
    } catch (Exception e) {
      e.printStackTrace();
```

---

* Sample Code: https://github.com/Baobaojianqiang/Min18Fragment3

```
    }
    Resources superRes = super.getResources();
    mResources = new Resources(mAssetManager,
    superRes.getDisplayMetrics(), superRes.
    getConfiguration());
    mTheme = mResources.newTheme();
    mTheme.setTo(super.getTheme());
  }

  @Override
  public AssetManager getAssets() {
    return mAssetManager == null ? super.getAssets()
    : mAssetManager;
  }

  @Override
  public Resources getResources() {
    return mResources == null ? super.getResources()
    : mResources;
  }

  @Override
  public Theme getTheme() {
    return mTheme == null ? super.getTheme() : mTheme;
  }
}
```

In the method *onCreate()* of *FragmentLoaderActivity*, we invoke the methods *loadClassLoader()* and *loadResources()*, and then load the *Fragment* of the plug-in.

```
public class FragmentLoaderActivity extends
BaseHostActivity {

  private String mClass;

  @Override
  protected void onCreate(Bundle savedInstanceState) {
    mDexPath = getIntent().getStringExtra(AppConstants.
    EXTRA_DEX_PATH);
    mClass = getIntent().getStringExtra(AppConstants.
    EXTRA_CLASS);
```

```
super.onCreate(savedInstanceState);
setContentView(R.layout.
activity_fragment_loader);
loadClassLoader();
loadResources();
try {
    //Reflects the Plug-In 's Fragment object
    Class<?> localClass = dexClassLoader.
    loadClass(mClass);
    Constructor<?> localConstructor =
    localClass.getConstructor(new Class[] {});
    Object instance = localConstructor.
    newInstance(new Object[] {});
    Fragment f = (Fragment) instance;
    FragmentManager fm = getFragmentManager();
    FragmentTransaction ft =
    fm.beginTransaction();
    ft.add(R.id.container, f);
    ft.commit();
} catch (Exception e) {
    Toast.makeText(this, e.getMessage(), Toast.
    LENGTH_LONG).show();
}
}
}
```

We introduce an ancient plug-in framework named *AndroidDynamic Loader* in this section. We can load the <u>Fragment</u> dynamically; we don't need to communicate with the *AMS* anymore.

10.3 DOWNGRADE*

If Google suddenly announced that Android plug-in technology was forbidden in all the Android app markets, such as it is with Google Play, then what would we do?

React Native may be the answer to this question, but if this technique is also forbidden, what would we do?

So back to the *Hybrid*? It's not a good solution; the performance of the mobile browser in Android is poor, especially on the list page.

---

* For the example code of this section, please refer to https://github.com/Baobaojianqiang/Hybrid1.2

Is there a mechanism that each *Activity* in the app has a corresponding *HTML5* page? When there is a bug or crash in one *Activity*, can we replace this *Activity* with an *HTML5* page immediately?

There are two problems to resolve here.

1) The original behavior is to click the button to jump from *ActivityA* to *ActivityB*, if we replace *ActivityA* with an *HTML* page named *page_a.html*, when we click the hyperlink in *page_a.html*, we expect the behavior to be the same as before, and to jump to *ActivityB*.

2) We find the navigation from *ActivityA* to *ActivityB* is as follows:

```
Intent intent = new Intent(MainActivity.this,
FirstActivity.class);
intent.putExtra("UserName", "jianqiang");
intent.putExtra("Age", 10);
startActivity(intent);
```

But we find the navigation from *ActivityA* to an *HTML* page named *page_a.html* is different. *ActivityA* can't navigate to an *HTML* page directly. We create *WebviewActivity* to load this *HTML5* page, and navigate from *ActivityA* to this *WebviewActivity*, shown as follows:

```
Intent intent = new Intent(MainActivity.this,
WebviewActivity.class);
newIntent.putExtra("FullURL",
  "file:///storage/emulated/0/myHTML5/thirdpage.
  html?a=1&b=abc");
startActivity(intent);
```

How to switch between *Activity* and *HTML5* freely without writing two different methods? The best solution is for the app developers to still write code in as usual; for example, they write the code "startActivity(intent)" to navigate the page from one *Activity* to another *Activity*. They don't care whether the next page is an *Activity* or HTML at runtime. We can override the methods *startActivity()* or *startActivityForResult()* of *Activity* to decide if the next page is an *Activity* or HTML. We need prepare a configuration file. This file specifies whether each page is an *Activity* or HTML. The app will download this file at runtime from the remote server. If the original Activity has a bug or crash, we can add a configuration in this file to specify which *HTML5* page will replace this *Activity*.

The solution has a name: downgrade. We don't need plug-in techniques to fix the online bugs anymore. The downgrade is also suitable for iOS. In the following sections, we talk about this technique in detail based on Android.

10.3.1 From *Activity* to *HTML5*

We hope that the Android developers start an *Activity* and pass data to this *Activity* as usual, including *Int*, *Strings*, and custom entities. They don't care about whether the next page is an *Activity* or an *HTML* page, as shown below:

```
Intent intent = new Intent(MainActivity.this,
FirstActivity.class);
intent.putExtra("UserName", "jianqiang");
intent.putExtra("Age", 10);

ArrayList<Course> courses = new ArrayList<Course>();
courses.add(new Course("Math", 80));
courses.add(new Course("English", 90));
courses.add(new Course("Chinese", 75));
intent.putExtra("Courses", courses);

startActivity(intent);
```

Then we download a *JSON* string from the remote server; it's a configuration file defining which activity will be replaced with an *HTML5* page, shown as follows:

```
[{
  "activity": "jianqiang.com.hook3.FirstActivity",
  "h5path": "file:///storage/emulated/0/myHTML5/
  firstpage.html",
  "fields": [{"fieldName": "UserName", type: 1},
    {"fieldName": "Age", type: 2},
    {"fieldName": "Courses", type: 3},
  ]
}, {
  "activity": "jianqiang.com.hook3.ThirdActivity",
  "h5path": "file:///storage/emulated/0/myHTML5/
  thirdpage.html"
}]
```

The field *fields* specifies the value of the parameters to navigate to *FirstActivity* or *firstpage.html*, as well as the type of the parameter, where one represents *String*, two represents *Int*, and three represents the custom entity. I don't list all the types in the demo; you can add more types as you need.

To be simple, we mock this configuration file in the *apk* and put the corresponding *HTML5* files in the folder *Assets* to simulate that we download these files from the remote server. The code is as follows:

```
public class MyApplication extends Application {
  public static HashMap<String, PageInfo> pages = new
  HashMap<String, PageInfo>();
}

  void prepareData() {
    String newFilePath = Environment.
    getExternalStorageDirectory() + File.separator +
    "myHTML5";
    Utils.copy(this, "firstpage.html", newFilePath);
    Utils.copy(this, "secondpage.html", newFilePath);
    Utils.copy(this, "thirdpage.html", newFilePath);
    Utils.copy(this, "style.css", newFilePath);

    String h5FilePath1 = newFilePath + File.separator
    + "firstpage.html";
    String h5FilePath2 = newFilePath + File.separator
    + "thirdpage.html";

    HashMap<String, Integer> fields = new
    HashMap<String, Integer>();
    fields.put("UserName", 1); //1 means string
    fields.put("Age", 2);     //2 means int
    fields.put("Courses", 3); //3 means object

    PageInfo pageInfo1 = new PageInfo("file://" +
    h5FilePath1, fields);
    MyApplication.pages.put("jianqiang.com.hook3.
    FirstActivity", pageInfo1);

    PageInfo pageInfo2 = new PageInfo("file://" +
    h5FilePath2, null);
    MyApplication.pages.put("jianqiang.com.hook3.
    ThirdActivity", pageInfo2);
  }
```

As the files are "downloaded from the remote server," we override the method *startActivityForResult()* of *BaseActivity* to parse this configuration file.

The methods *startActivity()* and *startActivityForResult()* are widely used in app development. *startActivity()* invokes *startActivityForResult ()*indirectly. That's why we override *startActivityForResult()* rather than *startActivity()*.

The logic of overriding the method *startActivityForResult()* is as follows:

1) Search in *MyApplication.pages* to check if we'll launch an *Activity* or an *HTML5* page. If we want to launch an *HTML5* page, we'll create a new *intent* to launch *WebViewActivity*, convert the parameters of the old *intent* to a string with the format "*url?k1=v1&k2=v2*". We need to encode *v1* and *v2* to a new string *str*, and pass the *URL* with the format "*URL?json=str*" to *WebViewActivity*.

We have downloaded a configuration file from the remote server. It's a *JSON* string. There is a field "*fields*" defined in this *JSON* string. We can use the corresponding Java syntax to take values from the *Intent* according to the type of each parameter defined in "*fields*," shown as follows:

```
public class BaseActivity extends Activity {
  @Override
  public void startActivityForResult(Intent intent,
  int requestCode) {
    if(intent.getComponent() == null) {
      super.startActivityForResult(intent,
      requestCode);
    }

    String originalTargetActivity = intent.
    getComponent().getClassName();

    PageInfo pageInfo = MyApplication.pages.
    get(originalTargetActivity);
    if(pageInfo == null) {
      super.startActivityForResult(intent,
      requestCode);
    }
```

```
StringBuilder sb2 = new StringBuilder();
if(pageInfo.getFields()!= null && pageInfo.
getFields().size() > 0) {
  sb2.append("{");

  for(String key: pageInfo.getFields().keySet()) {
    int type = pageInfo.getFields().get(key);
    switch (type) {
      case 1:
        String v1 = intent.getStringExtra(key);
        sb2.append("\"" + key + "\"");
        sb2.append(":");
        sb2.append("\"" + v1 + "\"");
        sb2.append(",");
        break;
      case 2:
        int v2 = intent.getIntExtra(key, 0);
        sb2.append("\"" + key + "\"");
        sb2.append(":");
        sb2.append(String.valueOf(v2));
        sb2.append(",");
        break;
      case 3:
        Serializable v3 = intent.
        getSerializableExtra(key);
        Gson gson = new Gson();
        String strJSON = gson.toJson(v3);
        sb2.append("\"" + key + "\"");
        sb2.append(":");
        sb2.append(strJSON);
        sb2.append(",");
        break;
      default:
        break;
    }
  }

  sb2.deleteCharAt(sb2.length() - 1);
  sb2.append("}");
}

StringBuilder sb = new StringBuilder();
sb.append(pageInfo.getUri());
```

```
if(pageInfo.getFields()!= null && pageInfo.
getFields().size() > 0) {
  sb.append("?json=");
  String str = null;
  try {
    str = URLEncoder.encode(sb2.toString(),
    "UTF-8");
  } catch (UnsupportedEncodingException e) {
    e.printStackTrace();
  }
  sb.append(str);
}

Intent newIntent = new Intent();
newIntent.putExtra("FullURL", sb.toString());

// The name of the alias' package, which is our
own package name.
String stubPackage = MyApplication.getContext().
getPackageName();

// Here we temporarily replace the started
Activity with WebviewActivity
ComponentName componentName = new
ComponentName(stubPackage, WebviewActivity.class.
getName());
newIntent.setComponent(componentName);

super.startActivityForResult(newIntent, requestCode);
  }
}
```

Let us go through the logic in *WebViewActivity*. It is used to receive the *JSON* data from the previous page and then pass this *JSON* data to the *WebView*.

```
public class WebviewActivity extends Activity {

  private static final String TAG = "WebviewActivity";
  WebView wv;
```

```
@Override
protected void onCreate(Bundle savedInstanceState) {
  super.onCreate(savedInstanceState);
  setContentView(R.layout.activity_webview);

  String fullURL = getIntent().getStringExtra
  ("FullURL");

  wv = (WebView) findViewById(R.id.wv);
  wv.getSettings().setJavaScriptEnabled(true);
  wv.getSettings().setBuiltInZoomControls(false);
  wv.getSettings().setSupportZoom(true);
  wv.getSettings().setUseWideViewPort(true);
  wv.getSettings().setLoadWithOverviewMode(true);
  wv.getSettings().setSupportMultipleWindows(true);
  wv.setWebViewClient(new MyWebChromeClient());
  wv.loadUrl(fullURL);
  }
}
```

In *firstpage.html*, we need to parse the parameters from the previous page, shown as follows:

```
<html>
<head>
  <meta charset="utf-8">
  <meta name="viewport"
  content="initial-scale=1,maximum-scale=1,minimum-
  scale=1,user-scalable=no"/>

  <link href="style.css" rel="stylesheet" type="text/
  css">
  <script type="text/JavaScript">
    function parseJSON() {
      var url = window.location.href;
      var arr = url.split('=');

      var otest = document.getElementById("test");

      var newli = document.createElement("li");

      var result = JSON.parse(decodeURIComponent
      (arr[1]));
```

```
      newli.innerHTML = result.UserName;

      otest.insertBefore(newli, otest.childNodes[1]);
    }
  </script>
</head>
<body onload="parseJSON()">
  <ul id="test">
    <li></li>
  </ul>
<body>
<html>
```

In *HTML5*, *JSON.parse()* was used to parse *JSON* data, and we use the method *decodeURIComponent()* to decode the value from *WebviewActivity*. Now, the navigation from *Activity* to *HTML5* is finished.

10.3.2 From *HTML5* to *Activity*

Now let's study how to navigate from *HTML* to *Activity*.

Firstly, we need to define some contracts. For example, if the app navigates to *SecondActivity*, we need to define a contract as follows:

```
activity://jianqiang.com.hook3.SecondActivity?
json=encodeData
```

encodeData in the above contract is the data we want to pass to *SecondActivity*. When we click the hyperlink with the above contract, the app will navigate to *SecondActivity* with the parameter "*json=encodeData*".

Sometimes when we click the hyperlink in *HTML5*, we want to navigate to another *HTML5* page, we need define the contract as follows:

```
secondpage.html? ?json=encodeData
```

Now let's talk about *the* generation of *encodeData*. For example, we want to pass the following *JSON* string to *Activity*:

```
{"HotelId":14, "HotelName" = "Hotel111", "Rooms":[{"room
Type":"LargeBed", "price":100}, {"roomType":"DoubleBed",
"price":200}]}
```

Unfortunately, we don't know the type of each field in this *JSON* string. So, we need provide another *JSON* string to define the type for each field, shown as follows:

```
[{"key": "HotelId", "value": "2"}, {"key":
"HotelName", "value": "1"}, {"key": "Rooms",
"value":"jianqiang.com.hook3.entity.Course"}]
```

One is *String*, two is *Integer*, three is a custom entity. It is consistent with the type number defined in Section 10.3.1.

Now let's merge these two *JSON* strings into one string, and encode this string, shown as follows in *firstpage.html*:

```
<html>
<head>
  <meta charset="utf-8">
  <meta name="viewport"
  content="initial-scale=1,maximum-scale=1,minimum-
  scale=1,user-scalable=no"/>

  <link href="style.css" rel="stylesheet" type="text/
  css">
  <script type="text/JavaScript">
    function gotoSecondActivity() {
      var baseURL = "activity://jianqiang.com.hook3.
      SecondActivity";
      var jsonValue = "{'HotelId':14, 'HotelName' =
      'guoguo hotel', 'Rooms':[{'roomType':'king
      bedroom', 'price':100}, {'roomType':'double-bed
      room', 'price':200}]}";
      var jsonType = "[{'key':'HotelId', 'value':'2'},
      {'key':'HotelName', 'value':'1'}, {'key':'Rooms',
      'value':'jianqiang.com.hook3.entity.Course'}]";
      var finalJSON = "{'jsonValue'=" + jsonValue + ",
      'jsonType'=" + jsonType + "}";

      baseURL = baseURL + "?json=" + encodeURIComponent
      (finalJSON);

      location.href= baseURL;
    }
```

```
function gotoSecondActivityInWeb() {
  var baseURL = "secondpage.html";
  var jsonValue = "{'HotelId':14, 'HotelName' =
  'guoguo hotel', 'Rooms':[{'roomType':'king
  bedroom', 'price':100}, {'roomType':'double-bed
  room', 'price':200}]}";
  var jsonType = "[{'key':'HotelId', 'value':'2'},
  {'key':'HotelName', 'value':'1'}, {'key':'Rooms',
  'value':'jianqiang.com.hook3.entity.Course'}]";
  var finalJSON = "{'jsonValue'=" + jsonValue + ",
  'jsonType'=" + jsonType + "}";


  location.href= baseURL+ "?json=" +
  encodeURIComponent(finalJSON);
  }
</script>
</head>
<body onload="parseJSON()">
  <a href="JavaScript:void(0)" onclick="gotoSecondActi
  vity()">navigate to SecondActivity</a> <br/>
  <a href="JavaScript:void(0)" onclick="gotoSecondActi
  vityInWeb()">navigate to SecondPage</a>
<body>
<html>
```

In *thridpage.html*, we define a new contract format, which is used to invoke the method *startActivityForResult()* of *Activity*:

```
startActivityForResult://jianqiang.com.hook3.
MainActivity
```

The code in *thridpage.html*:

```
<html>
<head>
  <meta charset="utf-8">
  <meta name="viewport" content="initial-scale=1,
  maximum-scale=1,minimum-scale=1,user-scalable=no"/>

  <link href="style.css" rel="stylesheet" type="text/
  css">
```

```html
<script type="text/JavaScript">
  function backToMainActivity() {
    var baseURL = "startActivityForResult://
    jianqiang.com.hook3.MainActivity";
    var jsonValue = "{'score':14}";
    var jsonType = "[{'key':'score', 'value':'2'}]";
    var finalJSON = "{'jsonValue'=" + jsonValue + ",
    'jsonType'=" + jsonType + "}";

    baseURL = baseURL + "?json=" + encodeURIComponent
    (finalJSON);

    location.href= baseURL;
  }

</script>
</head>
<body>
  <ul id="test">
    <li></li>
  </ul>

  <a href="JavaScript:void(0)" onclick="backToMainActi
  vity()">return results</a> <br/>
<body>
<html>
```

In *WebViewActivity* which loads *firstpage.html*, we need intercept the
URL. We dispatch the request to the different methods of *Activity* using
the contract.

Let's have a look at the complete code of *WebViewActivity*:

```java
public class WebviewActivity extends Activity {

  private static final String TAG = "WebviewActivity";
  WebView wv;

  @Override
  protected void onCreate(Bundle savedInstanceState) {
    super.onCreate(savedInstanceState);
    setContentView(R.layout.activity_webview);
```

```
    String fullURL = getIntent().
    getStringExtra("FullURL");

    wv = (WebView) findViewById(R.id.wv);
    wv.getSettings().setJavaScriptEnabled(true);
    wv.getSettings().setBuiltInZoomControls(false);
    wv.getSettings().setSupportZoom(true);
    wv.getSettings().setUseWideViewPort(true);
    wv.getSettings().setLoadWithOverviewMode(true);
    wv.getSettings().setSupportMultipleWindows(true);
    wv.setWebViewClient(new MyWebChromeClient());
    wv.loadUrl(fullURL);
}

public class MyWebChromeClient extends WebViewClient {
  @Override
  public boolean shouldOverrideUrlLoading(WebView
  view, WebResourceRequest request) {
    Uri url = request.getUrl();

    if(url == null) {
      return super.shouldOverrideUrlLoading(view,
      request);
    }

    Intent intent = null;
    if (url.toString().toLowerCase().
    startsWith("activity://")) {
      intent = parseUrl(url.toString(), "activity://");
      startActivity(intent);
    } else if(url.toString().toLowerCase().startsWith
    ("startactivityforresult://")) {
      intent = parseUrl(url.toString(),
      "startactivityforresult://");
      setResult(2, intent);
      finish();
    } else {
      return super.shouldOverrideUrlLoading(view,
      request);
    }
```

```
    return true;
  }
}

Intent parseUrl(String url, String prefix) {
  int pos = url.indexOf("?");
  String activity = url.substring(prefix.length(),
  pos);

  //6 means ?json=
  String jsonEncodeData = url.substring(pos + 6);

  String jsonData = null;
  try {
    jsonData = URLDecoder.decode(jsonEncodeData,
    "UTF-8");
  } catch (UnsupportedEncodingException e) {
    e.printStackTrace();
  }

  JSONObject jsonObject = null;
  try {
    jsonObject = new JSONObject(jsonData);
  } catch (JSONException e) {
    e.printStackTrace();
  }

  JSONArray jsonType = jsonObject.
  optJSONArray("jsonType");
  JSONObject jsonValue = jsonObject.
  optJSONObject("jsonValue");

  Intent intent = new Intent();

  for (int i = 0; i < jsonType.length(); i++) {
    JSONObject item = jsonType.optJSONObject(i);
    String key = item.optString("key");
    String value = item.optString("value");

    switch (value) {
      case "1":
        String strData = jsonValue.optString(key);
        intent.putExtra(key, strData);
```

```
      break;
    case "2":
      int intData = jsonValue.optInt(key);
      intent.putExtra(key, intData);
      break;
    default:
      JSONArray arrayData = jsonValue.
      optJSONArray(key);
      Gson gson = new Gson();
      ArrayList arrayList = new ArrayList();

      try {
        for (int j = 0; j < arrayData.length(); j++) {
          Object data = gson.fromJson(arrayData.
          optJSONObject(j).toString(), Class.
          forName(value));
          arrayList.add(data);
        }
      } catch (ClassNotFoundException e) {
        e.printStackTrace();
      }

      intent.putExtra(key, arrayList);

      break;
  }
}

ComponentName componentName = new
ComponentName(getPackageName(), activity);
intent.setComponent(componentName);

return intent;
  }
}
```

10.3.3 Support for the Backpress Button

The Android system has a backpress button.

In the downgrade solution, we need to support the backpress button. When the app navigates from one *HTML5* page to another *HTML5* page, and we press the backpress button, we want the app to go back to the first HTML5 page; we need to override the method *onKeyDown()* of *WebViewActivity*:

```
@Override
public boolean onKeyDown(int keyCode, KeyEvent
event) {
  if (event.getAction() == KeyEvent.ACTION_DOWN) {
    if (keyCode == KeyEvent.KEYCODE_BACK &&
    wv.canGoBack()) { // Indicates the operation when
    the physical back button is pressed
      wv.goBack(); // back
      // webview.goForward();//forward
      return true; // processed
    }
  }
  return super.onKeyDown(keyCode, event);
}
```

Downgrade is a good solution for fixing online bugs of the app without publishing a new version. It's a temporary solution to fix bugs quickly. It supports both Android and iOS.

10.4 PROGUARD FOR PLUG-INS

Plug-ins can be signed.

This section introduces the techniques to obfuse the plug-in after the plug-in is signed.

10.4.1 Basic Obfuse Rules for Plug-Ins*

In Android, we use ProGuard to obfuse the Android code.

Let's have a look at what we need to consider when we obfuse a simple app.

- Four components (*Activity* and so on) and *Application* should be declared in the *AndroidManifest.xml*; they cannot be obfused.

- *R.java* cannot be obfused because sometimes we need to fetch resources of the app using reflection syntax.

---

* Code sample: https://github.com/Baobaojianqiang/Sign1 & https://github.com/Baobaojianqiang/ Sign2

- Class *in android.support.v4* and *android.support.v7* cannot be obfused.

- A class that implements *Serializable* cannot be obfused; otherwise, it will throw an exception in deserialize.

- Generics cannot be obfused.

- *CustomView* cannot be obfused; otherwise, we can't find this *CustomView* in the layout.

These rules also can be applied to plug-ins, because the plug-in is also an *apk*. Although we don't need to declare four components in the *AndroidManifest.xml* of the plug-in, we still get these components from the plug-in using the reflection of their full name.

Sometimes, the HostApp may invoke a method in the plug-in by reflection, shown as follows:

```
Class mLoadClass = classLoader.loadClass("jianqiang.
com.receivertest.MainActivity");
Object mainActivity = mLoadClass.newInstance();

Method getNameMethod = mLoadClass.
getMethod("doSomething");
getNameMethod.setAccessible(true);
String name = (String) getNameMethod.
invoke(mainActivity);
```

Run these codes above and the app will throw an exception that the method *doSomething()* of *MainActivity* is not found because it's obfused.

So, we could obfuse the classes or the methods invoked by reflection in Plug-In, shown as follows, we need to add a configuration in Proguard-rules:

```
-keep class jianqiang.com.receivertest.MainActivity {
  public void doSomething();
}
```

10.4.2 Obfuse Without a Common Library*

HostApp and the plug-in app always use the same class library, shown in Figure 10.6.

---

* Code sample: https://github.com/Baobaojianqiang/ZeusStudy1.5

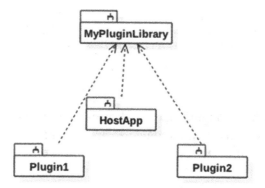

FIGURE 10.6   Plug-in architecture.

In Figure 10.6, the HostApp and plug-ins both have a dependency on a common library named MyPluginLibrary. If we compile MyPluginLibrary into the HostApp and plug-ins and package all the plug-ins into the HostApp, the HostApp will have at least two MyPluginLibraries, and the size of the HostApp will be larger. So, we use the keyword "provided" in the *build.gradle* of the plug-in and use the keyword "compile" in the *build.gradle* of the HostApp.

Is it necessary to obfuse *MyPluginLibrary*?

In this section, we focus on the plug-in solution without obfusing MyPluginLibrary; it's simple to implement.

1) Use the keyword "provided" in the *build.gradle* of Plugin1, shown as follows:

```
dependencies {
  compile fileTree(dir: "libs", include: ["*.jar"])
  testCompile "junit:junit:4.12"

  compile "com.android.support:appcompat-v7:25.2.0"

  //compile project(":MyPluginLibrary")
  provided files("lib/mypluginlibrary.jar")
}
```

2) Modify the *build.gradle* to obfuse Plugin1:

```
buildTypes {
  release {
```

```
minifyEnabled true
proguardFiles getDefaultProguardFile("proguard-
android.txt'", "proguard-rules.pro"
  }
}
```

Now we package Plugin1 again and open it in JadxGUI; we can find that the code is obfused and MyPluginLibrary is missing in *plugin1.apk*, shown in Figure 10.7.

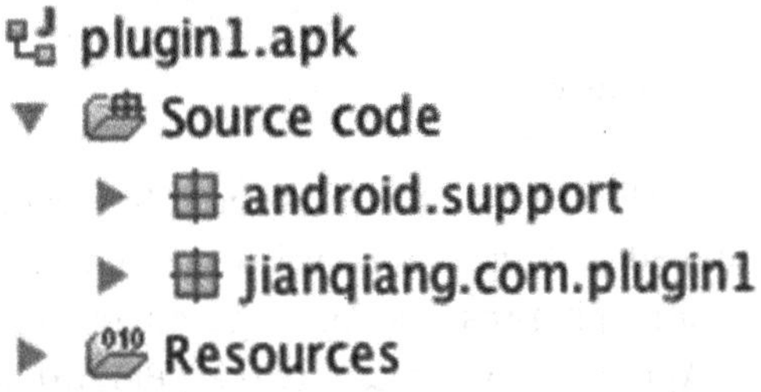

plugin1.apk
▼ **Source code**
  ▶ **android.support**
  ▶ **jianqiang.com.plugin1**
▶ **Resources**

FIGURE 10.7   Structure of Plugin1 without MyPluginLibrary.

Because MyPluginLibrary is not packaged into *plugin1.apk*, it's not obfused. Let's check the *TestActivity1* of Plugin1 in Figure 10.7; we find it uses *PluginManager* of MyPluginLibrary as usual, shown as follows:

```
public class TestActivity1 extends ZeusBaseActivity {
  protected void onCreate(Bundle bundle) {
    super.onCreate(bundle);
    setContentView(R.layout.activity_test1);
    findViewById(R.id.btnGotoActivityA).
    setOnClickListener(new OnClickListener(this) {
      final /* synthetic */ TestActivity1 a;
      {
        this.a = r1;
      }

      public void onClick(View view) {
        try {
          Intent intent = new Intent();
          intent.setComponent(new
          ComponentName("jianqiang.com.plugin1",
          "jianqiang.com.plugin1.ActivityA"));
```

```
        intent.putExtra("UserInfo", new
        UserInfo("baobao", 60));
        intent.putExtra("PlugPath", ((PluginItem)
        PluginManager.plugins.get(0)).pluginPath);
        this.a.startActivity(intent);
      } catch (Exception e) {
        e.printStackTrace();
      }
    }
  });
  }
}
```

Because MyPluginLibrary is packaged into the HostApp, the HostApp should follow the same rule as Plugin1, and not obfuse MyPluginLibrary either; otherwise, when we invoke the class of MyPluginLibrary, it will throw a *ClassNotFoundException*. We need to add a rule to avoid obfusing any class of MyPluginLibrary, shown as follows:

```
-keep class com.example.jianqiang.mypluginlibrary.** {
*;}
```

In the next section, we focus on how to obfuse MyPluginLibrary with a plug-in solution.

10.4.3 Obfusing with a Common Library*

I have introduced a simple plug-in solution without obfusing the common library in Section 10.4.2. It's simple, but not secure. If we put some important logic into the common library without obfusing, it's easy to get the source code of the common library by JadxGUI.

So, it's necessary to obfuse the common library. We still take the HostApp, Plugin1, and *MyPluginLibrary* as an example. The HostApp and plug-ins must obfuse MyPluginLibrary with the same rule.

If we obfuse Plugin1, we need to use the keyword "compile" rather than "provided" in the *build.gradle* of Plugin1. But the keyword "compile" will increase the size of the final *apk*; we have talked about this in Section 10.4.3.

A perfect solution should include the following three points:

---

* Code sample: https://github.com/Baobaojianqiang/ZeusStudy1.6

1) We find that we can split the original *dex* of Plugin1 into two *dexes*, *classes.dex* and *classes2.dex*. The classes of Plugin1 are thrown into *classes.dex*, the other classes are thrown into *classes2.dex*, including MyPluginLibrary. We can use MultiDex to finish this work.

2) And then we replace *classes2.dex* with an empty *dex* with the same name manually. We need to write a Python script to finish this work.

3) Finally, we use the same obfusing rule in the HostApp and Plugin1. We obfuse Plugin1 at first and then put the obfusing rule defined in Plugin1 into the HostApp. The HostApp will follow the same obfusing rule as Plugin1.

I will introduce the three aspects in the following section.

10.4.3.1 Use MultiDex

First, we add a configuration in the *build.gradle* of Plugin1 to support MultiDex as follows:

```
dexOptions {
  javaMaxHeapSize "4g"
  preDexLibraries = false

  additionalParameters += '--multi-dex'
  additionalParameters += '--main-dex-
  list=maindexlist.txt'
  additionalParameters += '--minimal-main-dex'
  additionalParameters += '--set-max-idx-number=20000'
}
```

We use the keyword "compile" in the *build.gradle* of Plugin1. MyPluginLibrary will be packaged into *plugin1.apk*:

```
compile project(path: ":MyPluginLibrary")
```

It's time to create a *maindexlist.txt* file, and this file contains all the classes packaged in *classes.dex*; we define all the classes of Plugin1 in this file, as follows:

```
jianqiang/com/plugin1/TestService1.class
jianqiang/com/plugin1/ActivityA.class
jianqiang/com/plugin1/TestActivity1.class
```

The classes which not defined in this file will be packaged into *classes2. dex*, such as the classes of *MyPluginLibrary*.

It is difficult to write these classes one by one in *maindexlist.txt*, so we write a Python script to generate all the classes. The code of *collect.py* is shown as follows:

```python
import os

fw = open('maindexlist.txt', 'w')

def dirlist(path):
  filelist = os.listdir(path)

  for filename in filelist:
    filepath = os.path.join(path, filename)
    if os.path.isdir(filepath):
      dirlist(filepath)
    elif len(filepath)>5 and filepath[-5:]=='.java':
      baseStr = filepath.replace('src/main/java/','').
      replace('.java', '')
      fw.write(baseStr+ '.class\n')
      for index in range(1, 11):
        fw.write(baseStr+ '$' + str(index) + '.class\n')
  fw.close()
dirlist("src/main/java/")
```

When we execute this script, it will scan all the files with the suffix "*.java*" in the project Plugin1 and replace this suffix "*.java*" with "*.class*".

But we find that there are many inner classes in Plugin1. For example, when we write code as follows, it will generate *ActivityA$1*, *ActivityB$2*, and so on.:

```java
Button b1 = new Button(this);
b1.setOnClickListener(new View.OnClickListener() {
  @Override
  public void onClick(View v) {
    //do something
  }
});
```

In the original Python script, it will generate a class like *ActivityA* without its inner classes. So we need to consider how to generate these inner classes automatically. We find it's impossible to know how many inner classes there are in a class in advance. We also find that MultiDex does not support the wildcard *. We can't use * in *maindexlist.txt*.

A simple but rough method is to generate 100 inner classes for each class in advance from ActivityA$1 to ActivityA$100. It's enough to use.

Let's modify *collect.py* to support inner classes.

```
import os

fw = open('maindexlist.txt', 'w')

def dirlist(path):
  filelist = os.listdir(path)

  for filename in filelist:
    filepath = os.path.join(path, filename)
    if os.path.isdir(filepath):
      dirlist(filepath)
    elif len(filepath)>5 and filepath[-5:]=='.java':
      baseStr = filepath.replace('src/main/java/','').
      replace('.java', '')
fw.write(baseStr+ '.class\n')
#generate nested class
      for index in range(1, 11):
        fw.write(baseStr+ '$' + str(index) + '.class\n')
  fw.close()
dirlist("src/main/java/")
```

Let's compile and package Plugin1; we will find that the original *dex* was separated into two *dexes*. *classes.dex* has all the classes of Plugin1, and the classes of *MyPluginLibrary* are thrown into *classes2.dex*. *classes2.dex* also has the classes of *android.support,** shown in Figures 10.8 and 10.9.

> 🖧 classes.dex
> ▼ 🗂 Source code
>   ▼ ⊞ jianqiang.com.plugin1
>   ▶ ⓖ ActivityA
>   ▶ ⓖ TestActivity1
>   ▶ ⓖ TestService1

FIGURE 10.8 *classes.dex* in JadxGUI.

- 🔲 classes2.dex
- ▼ 📦 Source code
  - ▼ ⊞ android.support
    - ▶ ⊞ annotation
    - ▶ ⊞ compat
    - ▶ ⊞ coreui
    - ▶ ⊞ coreutils
    - ▶ ⊞ fragment
    - ▶ ⊞ graphics.drawable
    - ▶ ⊞ mediacompat
    - ▶ ⊞ v4
    - ▶ ⊞ v7
  - ▼ ⊞ com.example.jianqiang.mypluginlibrary
    - ▶ ⓖ UserInfo
    - ▶ ⓖ ZeusBaseActivity
    - ▶ ⓖ a
    - ▶ ⓖ b

FIGURE 10.9 *classes2.dex* in JadxGUI.

10.4.3.2 Modify the ProGuard File

Suppose there are five classes, A, B, C, D, and E in MyPluginLibrary, the HostApp uses A, B, and C, and Plugin1 uses C, D, E. When we obfuse the HostApp because D and E are not used in the HostApp, it will be removed from MyPluginLibrary. When we run the logic of Plugin1, it will throw a *ClassNotFoundException* because D and E are missing.

To resolve this bug, we need to add a configuration in *proguard-rule.pro* of the HostApp and Plugin1, shown as follows:

```
-dontshrink
```

It means all the classes are kept in obfusing; even if they are not used in the current app.

When we obfuse Plugin1, it will generate a mapping file in the directory *Build/Output/Mapping/Release*. If the name of the class is obfused, the original name of the new name of this class will be stored in this file. It forms a mapping. This file maintains all the mappings of the original name to the new name. This is shown as follows:

```
com.example.jianqiang.mypluginlibrary.AppConstants ->
com.example.jianqiang.mypluginlibrary.a:
  java.lang.String PROXY_VIEW_ACTION -> a
```

```
java.lang.String EXTRA_DEX_PATH -> b
java.lang.String EXTRA_CLASS -> c
6:6:void <init>() -> <init>
com.example.jianqiang.mypluginlibrary.BuildConfig ->
com.example.jianqiang.mypluginlibrary.b:
  boolean DEBUG -> a
  java.lang.String APPLICATION_ID -> b
  java.lang.String BUILD_TYPE -> c
  java.lang.String FLAVOR -> d
  int VERSION_CODE -> e
  java.lang.String VERSION_NAME -> f
6:6:void <init>() -> <init>
```

Copy these mappings and save it as a file named *mapping_mypluginlibrary.txt*. Then copy this mapping file into the root directory of the HostApp and add a configuration in *proguard-rules.pro* of the HostApp:

```
-applymapping mapping_mypluginlibrary.txt
```

Up until now, the HostApp will apply the same obfusing rules as Plugin1.

10.4.3.3 Remove Redundant *Dexes from* plugin1.apk

Because MyPluginLibrary is compiled into *plugin1.apk*, the size of this *apk* file is larger. Not only MyPluginLibrary but also a lot of redundant libraries are packaged into *plugin1.apk*, such as android.support.*

It's time to reduce the size of the app. Let's prepare something as follows:

- The *keystore.jks* file of the HostApp and Plugin1, which is the private key to sign the HostApp and Plugin1.

- Create a script file named *createEmplyDex.py*.

- Package Plugin1 to generate *plugin1.apk*.

We split the *dex* of *plugin1.apk* into two *dexes* in Section 10.4.3.2. All the classes of Plugin1 are thrown into *classes.dex*. The other classes are thrown into *classes2.dex*, including the classes of *MyPluginLibrary*.

Now let's perform our magic tricks.

1) Write a python script to create an empty *dex*, shown as follows:

```
f=open('classes2.dex','w')
f.close()
```

2) Use *apktool* to decompile *plugin1.apk*.

```
java -jar apktool.jar d --no-src -f plugin1.apk
```

It will generate a subdirectory *Plugin1* in the current directory; there are two *dex* files in this directory, shown in Figure 10.10.

AndroidManifest.xml
apktool.yml
build
classes.dex
classes2.dex
dist
original
res

FIGURE 10.10   Structure of *plugin1.apk*.

3) Replace *classes2.dex* with the empty *dex* we created in step 1.

4) Package and generate *plugin1.apk* again, using the following command:

```
java -jar apktool.jar b plugin1
```

5) Sign *plugin1.apk* again with the following command, key0 is the alias of *keystore.jks* and the password is 123456.

```
jarsigner -verbose -keystore keystore.jks -digestalg
SHA1 -sigalg MD5withRSA -signed jar plugin1_sign.apk
"plugin1/dist/plugin1.apk" key0
```

6) Execute the command zipalign again, shown as follows:

```
zipalign -v four plugin1_sign.apk plugin1_final.apk
```

The original size of *plugin1.apk* was 1.4M; after these six steps, the size of *plugin1_final.apk* is only 620k. Finally, we rename *plugin1_final.apk* to *plugin1.apk* and put it in the folder *Assets* of the HostApp.

To obfuse Android code is boring and tedious, especially the second solution introduced in Section 10.4.3. We can merge all these steps into Gradle for simplicity,* but it will be difficult to understand its internal principle.

10.5 INCREMENTAL UPDATE†

In this section, let's talk about how to update the plug-in, which is an important feature. Each plug-in is an app; the size of the app is large. When the version of the plug-in upgrades from 1.0 to 2.0, the app will download the new plug-in from the remote server. If the size of this plug-in is larger than 10M, so it will take a long time.

If we can generate a file that includes only the differences between the versions 1.0 and 2.0 of this plug-in and supply this file for the App user, the size of this file is small; the app user can download it in a short time.

This technique has a cool name: An Incremental Update.

10.5.1 The Basic Concept of an Incremental Update

First, we need to compile and package all the plug-ins and put these plug-ins in the folder *Assets* of the HostApp, to make sure the HostApp can load all the plug-ins in this folder.

Each plug-in has its own version. If the version of the HostApp is 6.0.0, the version of the plug-in in the folder *Assets* of the HostApp should be 6.0.0.1.

When we find that Plugin1 version 6.0.0.1 has a bug, we fix this bug, compile and package Plugin1 again, and set its version to 6.0.0.2, and put it on the remote server. The app will detect the new version of Plugin1 and download it.

---

* Refer to https://github.com/louiszgm/zeusstudy1.6
† Sample Code: https://github.com/That2

If the size of Plugin1 is large, for example, more than 30M, and it will take a long time to download it. But we find the differences between the new plug-in and the old plug-in are small, maybe less than 1M, so we use an incremental update technique to download these differences in a short time, and merge them with the old plug-in, to generate the new version of the plug-in.

10.5.2 Create an Incremental Package

We use *bsdiff* to compare these two versions of Plugin1, and to generate a file *patch.diff*.

Download the tool *bsdiff* and execute the following command to generate an Incremental Package *mypatch.diff*:

```
bspatch old.apk new.apk mypatch.diff
```

Zip the file *mypatch.diff*. It's better to generate this zip file on a Windows platform.

Upload this zip package, called *patch1.zip*, to the remote server. For example:

https://files.cnblogs.com/files/Jax/patch1.zip.

10.5.3 Apply Permissions

It's easy to download and unzip the Incremental Package; so, we won't spend too much time on it, please refer to the methods *download()* and *unzip()* in detail.

These two methods both need to read and write SDCard permissions. We need to write code to apply these permissions, shown as follows:

```
private static final int REQUEST_EXTERNAL_STORAGE = 1;
private static String[] PERMISSIONS_STORAGE = {
  Manifest.permission.READ_EXTERNAL_STORAGE,
  Manifest.permission.WRITE_EXTERNAL_STORAGE
};

public void verifyStoragePermissions() {
  // Check if we have write permission
  int permission = ActivityCompat.
  checkSelfPermission(this,
    Manifest.permission.WRITE_EXTERNAL_STORAGE);

If (permission != PackageManager.PERMISSION_GRANTED) {
    // We don't have permission so prompt the user
```

```
ActivityCompat.requestPermissions(this,
PERMISSIONS_STORAGE,
REQUEST_EXTERNAL_STORAGE);
    }
}
```

In the method *onCreate()* of *Activity*, we execute the method *verifyStoragePermissions()* to apply the permissions to read and write on to an SDCard. After the app launches, a dialog will pop up for the app user to confirm.

10.5.4 Merge Incremental Package

Now I introduce an open source project, *ApkPatchLibrary*, written by Cundong Liu*. It helps us to build and apply patches to binary files. *ApkPatchLibrary* is based on another famous open source project, *bsdiff*, but *ApkPatchLibrary* is used in the Android system. It supplies a tool named *libApkPatchLibrary.so*. We can use this tool to generate a file, including the differences between the two files; we named it *mypatch.diff*, we can also use this tool to merge the old file and *mypatch.diff* to generate the new file.

Now let's put *libApkPatchLibrary.so* in the folder *Jnilibs/Armeabi*, and then we create a package *com.cundong.utils* in the HostApp and create a java file *PatchUtils* in this package, as shown in Figure 10.11.

FIGURE 10.11   Incrementally updated SO packages.

---

* Refer to https://github.com/cundong/SmartAppUpdates

```
package com.cundong.utils;
public class PatchUtils {

/**
* native method Use the App with the path oldApkPath
and the patch with the path patchPath to generate a
new Apk and stores it in newApkPath.
*
* Returns 0, indicating successful operation
*
* @param oldApkPath example: /sdcard/old.apk
* @param newApkPath example: /sdcard/new.apk
* @param patchPath example: /sdcard/xx.patch
* @return
*/
public static native int patch(String oldApkPath,
String newApkPath,
String patchPath);
}
```

Now we can merge the files using *PatchUtils*, as follows:

1) In *MainActivity*, add a static method to initialize *ApkPatchLibrary*.

```
Static {
  System.loadLibrary("ApkPatchLibrary");
}
```

2) Merge files after the method *unzip()*:

```
try {
  int patchResult = PatchUtils.patch(oldApkPath,
  newFilePath, patchFilePath);
  if(patchResult == 0) {
  log.e("bao", patchResult + "");
} catch (Exception ex) {
  ex.printStackTrace();
}
```

newFilePath is the path of the new *apk* file after merging the files. Let's read this file to get the new version of the plug-in.

Up until now, we have finished an example of a plug-in incremental update. It is only a simple example. In an enterprise app, an incremental update is more complex.

Incremental update not only applies to the Android plug-in technology but also applies in *Hybrid* and *React Native*.

10.6 A PLUG-IN SOLUTION FOR SO FILES

In this section, we will talk about SO files.

During Android app development, Java code is not suitable, for example, for encryption algorithms or encoding and decoding of audio and video. We must use C/C++ to implement these special scenarios, to generate a file with the file extension "so." We always call it as a SO file. In the Android system, we use *System.load()* or *System.loadLibrary()* to load a SO file.

App developers always use SO supplied by a third party; for example, in Section 10.5, we use *libApkPatchLibrary.so* for an incremental update.

App developers rarely write SO. So, let's start from how to write an SO in this section.

10.6.1 Write a Hello-World SO*

In this section, we'll write a simple SO file with only one method *getString()*, which returns a string directly.

10.6.1.1 Download NDK

First, we need to download the Android NDK in Android Studio. Click *SDK Manager* in Android Studio and switch to the tab "SDK Tools," and here we can find an NDK item. Click on it to begin downloading NDK, as shown in Figure 10.12.

After we download NDK, and we can check the path of NDK in Figure 10.13.

Now we can find a configuration file generated automatically in the file *local.properties* in the root directory of the project.

```
ndk.dir=/Users/jianqiang/Library/Android/sdk/
ndk-bundle
```

---

* For Demo Codes,Goto *https://github.com/Baobaojianqiang/JniHelloWorld*

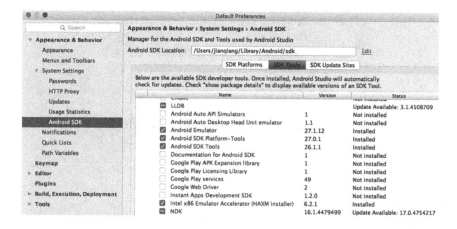

FIGURE 10.12   Download NDK.

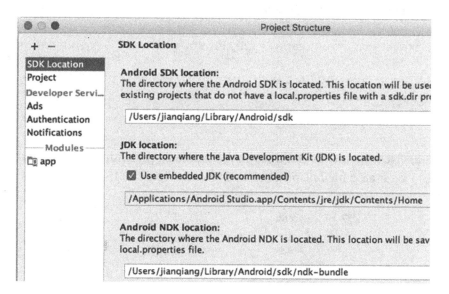

FIGURE 10.13   Configure NDK location.

Finally, we add a configuration in the file *gradle.properties.*

```
android.useDeprecatedNdk=true
```

10.6.1.2 Create a Project to Generate SO

Let's create an Android project; the project name is JniHelloWorld. Please follow me step by step:

FIGURE 10.14   Structure of JniHelloWorld.

1) Create a Java class named *JniUtils*, as shown in Figure 10.14, declare a method *getString()* in *JniUtils*.

```
package com.jianqiang.jnihelloworld;

public class JniUtils {
  public native String getString();
}
```

2) Click the menu "Build-> Make Module App." Android Studio will generate a file *JniUtils.class*, as shown in Figure 10.15.

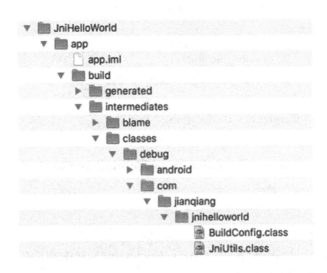

FIGURE 10.15   Location of *JniUtils.class*.

Go to the directory *Debug* and execute the following command:

```
javah -jni com.jianqiang.jnihelloworld.JniUtils
```

It will generate a file *com_jianqiang_jnihelloworld_JniUtils.h*; the content of this file is as follows:

```
/* DO NOT EDIT THIS FILE - it is machine generated */
#include <jni.h>
/* Header for class com_jianqiang_jnihelloworld_
JniUtils */

#ifndef _Included_com_jianqiang_jnihelloworld_JniUtils
#define _Included_com_jianqiang_jnihelloworld_JniUtils
#ifdef __cplusplus
extern "C" {
#endif
/*
 * Class:    com_jianqiang_jnihelloworld_JniUtils
 * Method:   getString
 * Signature: ()Ljava/lang/String;
 */
JNIEXPORT jstring JNICALL Java_com_jianqiang_
jnihelloworld_JniUtils_getString
  (JNIEnv *, jobject);

#ifdef __cplusplus
}
#endif
#endif
```

3) In the directory *Jinhelloworld/App/Src/Main*, create a folder named *Jni*, and copy the file *com_jianqiang_jnihelloworld_JniUtils.h* into this folder, as shown in Figure 10.16.

FIGURE 10.16   Location of *com_jianqiang_jnihelloworld_JniUtils.h*.

In the same folder *Jni*, create a file *com_jianqiang_jnihelloworld_JniUtils.c*; the content of this file is as follows:

```
#include "com_jianqiang_jnihelloworld_JniUtils.h"
/**
 * The content of the above include directive must
 have .h extension, and the following function must
 have the same name of this file
 */
JNIEXPORT jstring JNICALL Java_com_jianqiang_
jnihelloworld_JniUtils_getString
    (JNIEnv *env, jobject obj) {
  return (*env)->NewStringUTF(env, "Hello Jianqiang");
}
```

4) Add an NDK configuration in the *build.gradle* to generate two SO files which support arm-32bits and arm-64bits.

```
apply plugin: "com.android.application"

android {
  compileSdkVersion 26
  buildToolsVersion "27.0.3"
  defaultConfig {
    applicationId "com.jianqiang.jnihelloworld"
    minSdkVersion 22
    targetSdkVersion 26
    versionCode 1
    versionName "1.0"
    testInstrumentationRunner "android.support.test.
    runner.AndroidJUnitRunner"

    ndk{
      moduleName "hello" // name of generated SO
      abiFilters "armeabi-v7a", "arm64-v8a"
    }
  }
  buildTypes {
    release {
      minifyEnabled false
```

```
    proguardFiles getDefaultProguardFile('proguard-
    android.txt'), 'proguard-rules.pro'
  }
}
}
```

5) Rebuild the project; Android Studio will generate a new directory
   NDK in *Build/Intermediates* automatically, and there are two subdirectories *Armeabi-V7a* and *Arm64-V8a*, which stores the corresponding SO file, *libhello.so*, shown in Figure 10.17.

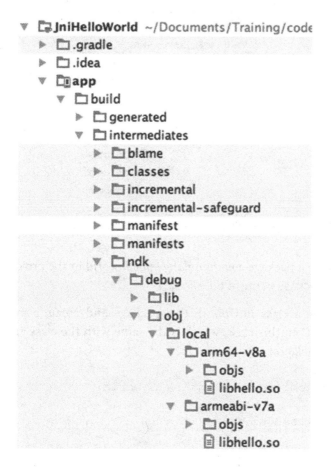

FIGURE 10.17   Directory of SO.

10.6.2 Using SO*

In this section, we'll try to use the *libhello.so* created in Section 10.6.1.

1) Create a folder named *Jnilibs*, then copy *libhello.so* to this folder, as shown in Figure 10.18.

FIGURE 10.18   Copy *libhello.so* to the folder *Jnilibs*.

2) Create a package *com.jianqiang.jnihelloworld* in the project; it's also the package name of *libhello.so*.

3) Define a class *JniUtils* in this package and create a method *getString()* in this class, which is the same with the class and method in *libhello.so*:

```
package com.jianqiang.jnihelloworld;

public class JniUtils {

  static {
    System.loadLibrary("hello");
  }
```

---

* For demo codes,Goto *https://github.com/Baobaojianqiang/MySO1*

```
//It must be declare using native when calling c
functions from Java and keep the names of functions
identical.
  public native String getString();
}
```

JniUtils has a static constructor. We always load *libhello.so* in the static constructor. Because the static constructor is executed only once to make sure this SO file is loaded only once too.

4) Finally, click the button "MainActivity," to invoke the method *getString()* of *libhello.so*:

```
public class MainActivity extends Activity {

  @Override
  protected void onCreate(Bundle savedInstanceState) {
    super.onCreate(savedInstanceState);
    setContentView(R.layout.activity_main);

    final Button btnShowMessage = (Button)
    findViewById(R.id.btnShowMessage);
    btnShowMessage.setOnClickListener(new View.
    OnClickListener() {
      @Override
      public void onClick(View v) {
        btnShowMessage.setText(new JniUtils().
        getString());
      }
    });
  }
}
```

Up until now, we finish a simple demo to illustrate how to SO. When the app is starting, the Android system will choose the corresponding SO in the folder *Arm64-V8a* or *Armeabi-V7a*.

10.6.3 The Principle of Loading SO

In Section 10.6.2, we created an SO file and used this SO file in another project. I will talk about the process of loading an SO file in this section.

10.6.3.1 Compiling SO*

Android supports three types of CPU: x86-based, ARM-based, and MIPS-based. In fact, there are a few devices based on x86 or mips. Nearly all the Android devices use ARM.

Now we only talk about ARM, which is divided into 32-bits and 64-bits:

- armeabi/armeabi-v7a: This is mainly used on Android 4.0 and higher versions. The CPU of the phone must be 32-bits. However, armeabi is a very old version; it is obsolete.

- arm64-v8: This is mainly used for Android 5.0 and higher versions. The CPU of the phone must be 64-bits.

Usually, we generate multiple SO files according to the CPU types and put them under the folder *Jnilibs* of the project, shown in Figure 10.19.

FIGURE 10.19 *Jnilibs*, containing different SO files.

Actually, we don't need to prepare so many SO files. ARM is downward-compatible; for example, SO compiled in 32-bits can be run normally on the 64-bits system.

Android starts one virtual machine (VM) for each app. When the Android 64-bits system loads 32-bits SO or app, it will start two VM; one is 64-bits, the other one is 32-bits.

So, it is enough to keep only one SO file to support armeabi-v7a in an app. Android 64-bit systems will load it on the 32-bit VM. It reduces the size of the app. Refer to the demo MySO1.1 in detail.

---

* For demo codes, refer to *https://github.com/Baobaojianqiang/MySO1.1*

10.6.3.2 The Process of Loading SO

Let's have a look at the whole process of loading the SO.

If the CPU is 64-bits, we can use the following command to check which CPUs the current phone supports. Take my phone; for example, it supports three types of SO, arm64-v8a, armeabi-v7a, and armeabi (Figure 10.20).

```
[baobao:~ jianqiang$ adb shell
[helium:/ $ getprop ro.product.cpu.abilist
arm64-v8a,armeabi-v7a,armeabi
```

FIGURE 10.20 Check how many types of SO files are supported in the current mobile phone.

All these types are stored in a sorted collection named *abiList*.

Iterate the directory *Jnilibs* in sequence. For example, the first one is arm64-v8a. If there is a subfolder named *Arm64-V8a* in the folder *Jnilibs* and there are SO files in the subfolder *Arm64-V8a*, then the Android system will load all the SO files in the subfolder *Arm64-V8a*.

For example, on my Android phone, there is an SO file *a.so* in the subfolder *Arm64-V8a*, and there are two SO files *a.so* and *b.so* in the subfolder *Armeabi-V7a*; my Android phone will load *a.so* in the subfolder *Arm64-V8a*, but the phone won't load *b.so* forever. It depends on the SO loading mechanism in the Android system.

To reduce the app size, we keep only one SO file, which is in the sub folder *Armeabi-V7a*.

32-bits ARM phones must load the SO files in armeabi-v7a.

If we want to load 32-bits SO file in a 64-bits ARM phone, we must put all the SO files in the subfolder *Armeabi-V7a*. We can't put any SO files in the subfolder *Arm64-V8a*.

10.6.3.3 Two Ways to Load SO*

There are two methods to load SO:

One way is to use *System.loadLibrary()*. It loads SO in the folder *Jnilibs*. For example, if we want to load *libhello.so*, we should write the following code:

```
System.loadLibrary("hello");
```

Another way is to use *System.load()*. It can load SO in any location. The parameter of this method is the full path of the SO file.

---

* For demo codes, go to https://github.com/Baobaojianqiang/MySO2

These two ways both use *dlopen* to open SO files.

We've talked about how to use *System.loadLibrary()* in the demo MySO1.

We can also put the SO files in the remote server. Once an app downloads these SO files, it will load them dynamically using the method *System.load()*. Refer to the demo MySO2.

Sometimes we only generate a 32-bits SO file and put it in the folder *Assets/Armeabi-V7a*.

Now if a 64-bits Android phone uses *System.load()* to load this SO file, it will throw an exception, shown as follows:

```
dlopen failed: libhello.so is 32-bit instead of 64 bits
```

Because this phone is a 64-bits VM, it can only load the 64-bits SO file. It can't load the 32-bits SO file.

In this scenario, we need write a simple 32-bits SO file and put this SO file in the folder *Jnilibs/Armeabi-V7* as a placeholder. The 64-bits Android system will use the 32-bits VM, rather than the 64-bits VM. So, the 64-bits Android system can load any SO file in the folder *Assets*.

For example, we write an SO file *libgoodbye.so*, which has only one method *sayGoodBye()* and returns a simple string "Goodbye baobao" directly. Refer to the demo JniHelloWorld2 for detail.

Out *libgoodbye.so* in the folder *Jnilibs/Armeabi-V7a* of MySO2, and then use *System.loadLibrary()* to load this SO file, and then use *System.load()* to load *libhello.so*. No exception occurs. The code is shown as follows:

```
public class MainActivity extends Activity {
  private String soFileName = "libhello.so";

  @Override
  protected void attachBaseContext(Context newBase) {
    super.attachBaseContext(newBase);

    Utils.extractAssets(newBase, soFileName);
  }

  @Override
  protected void onCreate(Bundle savedInstanceState) {
```

```java
super.onCreate(savedInstanceState);
setContentView(R.layout.activity_main);

File dir = this.getDir("jniLibs", Activity.
MODE_PRIVATE);

File tmpFile = new File(dir.getAbsolutePath() +
File.separator + "armeabi-v7a");
if (!tmpFile.exists()) {
  tmpFile.mkdirs();
}

File distFile = new File(tmpFile.getAbsolutePath()
+ File.separator + "libhello.so");

System.loadLibrary("goodbye");

if (Utils.copyFileFromAssets(this, "libhello.so",
distFile.getAbsolutePath())){
  System.load(distFile.getAbsolutePath());
}

final Button btnShowMessage = (Button)
findViewById(R.id.btnShowMessage);
btnShowMessage.setOnClickListener(new View.
OnClickListener() {
  @Override
  public void onClick(View v) {
    btnShowMessage.setText(new JniUtils().
    getString());
  }
});
  }
}
```

Loading SO dynamically is a good solution. If we don't need to load SO immediately, we can put all the SO files in the remote server to reduce the size of the *apk*.

On the other hand, we only need one SO file in the subfolder *Armeabi-V7a* to make the size of the *apk* smaller.

10.6.3.4 The Relationship between ClassLoader *and SO*

In Chapter 6, we used *DexClassLoader* to load classes of plug-ins, shown as follows:

```
File extractFile = this.getFileStreamPath(apkName);
dexpath = extractFile.getPath();

fileRelease = getDir("dex", 0); //0 For Context.
MODE_PRIVATE

classLoader = new DexClassLoader(dexpath,
    fileRelease.getAbsolutePath(), null,
    getClassLoader());
```

We always set the third parameter of the constructor of *DexClassLoader* to *null*. This parameter is the absolute path of the SO file.

10.6.4 A Plug-In Solution Based on *System.load()**

There are two solutions for downloading SO dynamically. This section will introduce the first solution.

In the first plug-in solution, the HostApp creates a *DexClassLoader* for each plug-in to parse classes of the plug-in. We can also get the path of each plug-in. We join these paths into a new string separated by commas. We can set this new string as the third parameter of the constructor of *DexClassLoader*. So, all the SO files in the HostApp and plug-in will "enjoy the same treatment." We use the method *System.loadLibrary()* to load these SO files.

Figure 10.21 shows the project structure of MySO3.

We use only one 32-bits SO file named *libgoodbye.so* in our project to reduce the size of the *apk*; put this file in the folder *Jnilibs*.

In the *UPFApplication* of the HostApp, we write code to load this 32-bits SO file to make sure the Android system uses the 32-bits virtual machine to load the other 32-bits SO file from now on.

The code is shown as follows:

```
public class UPFApplication extends Application {
  @Override
  protected void attachBaseContext(Context base) {
    super.attachBaseContext(base);
```

---

* For demo codes, go to *https://github.com/Baobaojianqiang/MySO3*

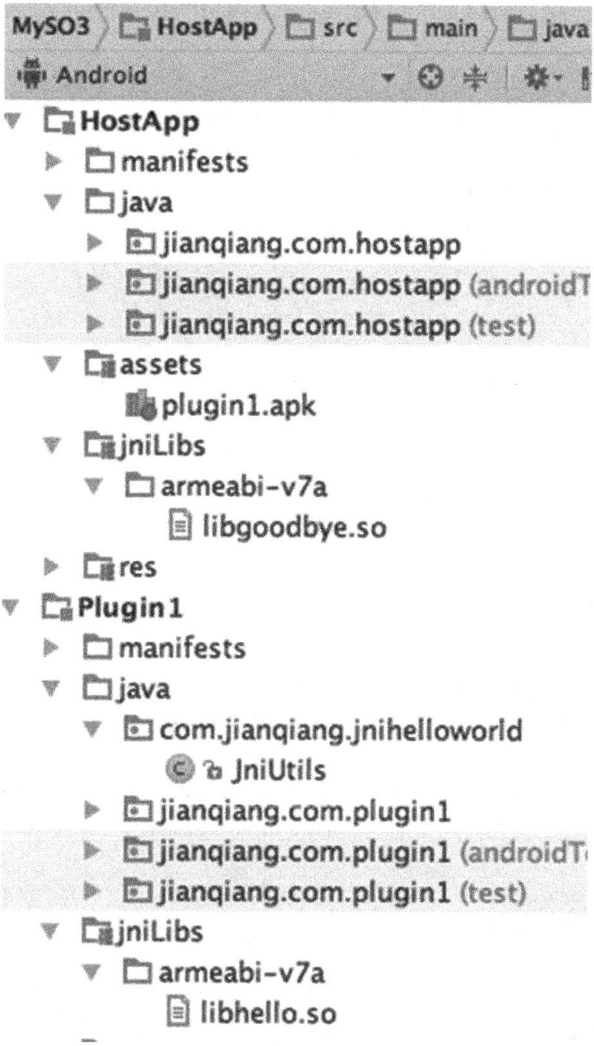

```
MySO3 〉 🗂 HostApp 〉 🗂 src 〉 🗂 main 〉 🗂 java
🤖 Android                    ▼ ⊙ ÷ | ✿▾
  ▼ 🗂 HostApp
    ▶ 🗂 manifests
    ▼ 🗂 java
      ▶ 🗂 jianqiang.com.hostapp
      ▶ 🗂 jianqiang.com.hostapp (androidT
      ▶ 🗂 jianqiang.com.hostapp (test)
    ▼ 🗂 assets
        🗃 plugin1.apk
    ▼ 🗂 jniLibs
      ▼ 🗂 armeabi-v7a
          📄 libgoodbye.so
    ▶ 🗂 res
  ▼ 🗂 Plugin1
    ▶ 🗂 manifests
    ▼ 🗂 java
      ▼ 🗂 com.jianqiang.jnihelloworld
          ©🅑 JniUtils
      ▶ 🗂 jianqiang.com.plugin1
      ▶ 🗂 jianqiang.com.plugin1 (androidT
      ▶ 🗂 jianqiang.com.plugin1 (test)
    ▼ 🗂 jniLibs
      ▼ 🗂 armeabi-v7a
          📄 libhello.so
```

FIGURE 10.21   Project structure of MySO3.

```
    System.loadLibrary("goodbye");
  }
}
```

The logic in the *MainActivity* of the HostApp:

```
public class MainActivity extends AppCompatActivity {

  private String dexpath = null;  //location of files
```

```java
private File fileRelease = null; //release folder
private DexClassLoader classLoader = null;

private String apkName = "plugin1.apk";  //apk name

TextView tv;

@Override
protected void attachBaseContext(Context newBase) {
  super.attachBaseContext(newBase);
  try {
    Utils.extractAssets(newBase, apkName);
  } catch (Throwable e) {
    e.printStackTrace();
  }
}

@SuppressLint("NewApi")
@Override
protected void onCreate(Bundle savedInstanceState) {
  super.onCreate(savedInstanceState);
  setContentView(R.layout.activity_main);

  File extractFile = this.getFileStreamPath(apkName);
  dexpath = extractFile.getPath();

  fileRelease = getDir("dex", 0); //0 For Context.
  MODE_PRIVATE

  //make a SO path concatenating with commas.
  String libPaths = Utils.UnzipSpecificFile(dexpath,
  extractFile.getParent());

  classLoader = new DexClassLoader(dexpath,
      fileRelease.getAbsolutePath(), libPaths,
      getClassLoader());

  tv = (TextView)findViewById(R.id.tv);
  Button btn_1 = (Button) findViewById(R.id.btn_1);
  // invoke with relection
  btn_1.setOnClickListener(new View.
  OnClickListener() {
```

```
@Override
public void onClick(View arg0) {
  Class mLoadClassBean;
  try {
    mLoadClassBean = classLoader.
    loadClass("jianqiang.com.plugin1.Bean");
    Object beanObject = mLoadClassBean.
    newInstance();

    Method getNameMethod = mLoadClassBean.
    getMethod("getName");
    getNameMethod.setAccessible(true);
    String name = (String) getNameMethod.
    invoke(beanObject);

    tv.setText(name);
    Toast.makeText(getApplicationContext(), name,
    Toast.LENGTH_LONG).show();

  } catch (Exception e) {
    Log.e("DEMO", "msg:" + e.getMessage());
  }
}
});
}
}
```

In Plugin1, we use the method *getName()* of *Bean* to invoke the method *getString()* defined in *libhello.so*, shown as follows:

```
public class Bean implements IBean {
  private String name = "jianqiang";

  private ICallback callback;

  @Override
  public String getName() {
    return new JniUtils().getString();
  }

  //ommit some codes
}
```

In Plugin1, we create a class *JniUtils* to load the SO file, shown as follows:

```
public class JniUtils {
  static {
    System.loadLibrary("hello");
  }

  //It must be declare using native when calling c
  functions from Java and keep the names of functions
  identical.
  public native String getString();
}
```

10.6.5 An SO Plug-In Solution Based on *System.loadLibrary()**

There are two solutions for downloading SO dynamically. This section will introduce the second solution.

The plug-in can load SO files by itself, copy SO files to a new folder, and use the method *System.loadLibrary()* to load them dynamically.

The solution is quite simple. I supply a demo base with ZeusStudy1.4 and modify two places, shown as follows:

1) Add a 32-bits SO file named *libgoodbye.so* to the folder *Jnilibs/ Armeabi-V7a* of the HostApp. Then invoke the method *System.load-Library()* in *Application* to load this SO file. Refer to Section 10.6.4 for details.

2) In *TestActivity1* of Plugin1, fetch *libhello.so* from the folder *Assets* of *plugin1.apk*, and then load it with the method *System.load()*, shown as follows:

```
public class TestActivity1 extends ZeusBaseActivity {
  private String apkName = "plugin1.apk";  //apk name
  private String soFileName = "libhello.so";

  @Override
  protected void attachBaseContext(Context newBase) {
    super.attachBaseContext(newBase);

    File extractFile = this.getFileStreamPath(apkName);
```

---

* For demo codes, go to https://github.com/Baobaojianqiang/ZeusStudy1.7

```
String dexpath = extractFile.getPath();

String libPath = Utils.UnzipSpecificFile(dexpath,
extractFile.getParent());

System.load(libPath + "/" + soFileName);
}

//ommit some lines of code.
}
```

We talk about the SO techniques in this section. We introduce two methods for loading SO files, which means we also have two solutions for loading the SO files of the plug-in.

10.7 HOOKING THE PACKAGING PROCESS

In this section, I will introduce a plug-in solution based on hooking the packaging process to resolve the conflict of resource ID. It has an interesting name: Small.* It is a custom Gradle plug-in, which modifies the *PackageId* of the resource ID.

10.7.1 Gradle Plug-In Project

Many Android plug-in frameworks write Gradle script to hook the packaging process to resolve the conflict of resource ID.

The Small framework provides a custom Gradle plug-in named Gradle-Small to hook *resources.arsc* during the packaging process.

Let's begin from how to write a custom Gradle plug-in.

10.7.1.1 Create Gradle Plug-In Project†

We can't create a Gradle plug-in project in Android Studio directly. We can create a Module or Android Library at first, and then delete all the files except *build.gradle* and the files in the folder *Src/Main*.

Next, we can use the project named buildSrc in code sample TestSmallGradle1.0, shown in Figure 10.22.

Now, in *build.gradle* of the buildSrc project, we add configuration as follows:

---

* https://github.com/wequick/Small
† For demo codes, go to https://github.com/Baobaojianqiang/TestSmallGradle1.0

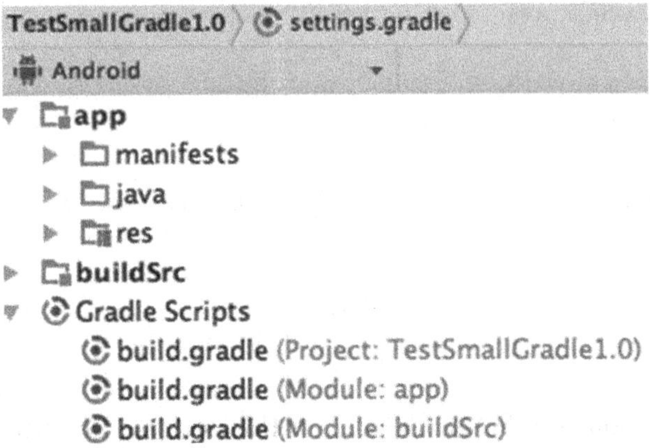

TestSmallGradle1.0 ⊙ settings.gradle

🤖 Android ▾

▼ 🗀 app
  ▶ 🗀 manifests
  ▶ 🗀 java
  ▶ 🗀 res
▶ 🗀 buildSrc
▼ ⊙ Gradle Scripts
  ⊙ build.gradle (Project: TestSmallGradle1.0)
  ⊙ build.gradle (Module: app)
  ⊙ build.gradle (Module: buildSrc)

FIGURE 10.22   Project structure of TestSmallGradle1.0.

```
apply plugin: "groovy"

dependencies {
  compile gradleApi()
  compile localGroovy()
}
```

And then we can create a class *MyPlugin.groovy* in the project build-Src. The grammar of "groovy" is the same as Java. *MyPlugin.groovy* implements the interface: plug-in<Project>, shown as follows:

```
public class MyPlugin implements plugin<Project> {

  @Override
  void apply(Project project) {
    project.task("testPlugin") << {
      println "Hello gradle plugin in src"
    }
  }
}
```

Then, let's create a sub folder in the folder *Src/Main*, shown in Figure 10.23.

In Figure 10.23, there is a file named *net.wequick.small.properties*; this is the entry point of the Gradle plug-in project. The content of this file is as follows; it will invoke the method *apply()* of *MyPlugin*:

```
implementation-class=com.jianqiang.MyPlugin
```

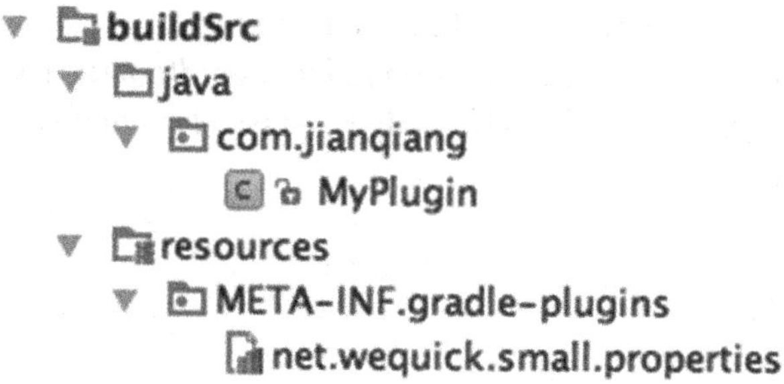

▼ 🗀 **buildSrc**
  ▼ 🗀 **java**
    ▼ 🗀 **com.jianqiang**
      🄲 🗋 **MyPlugin**
  ▼ 🗀 **resources**
    ▼ 🗀 **META-INF.gradle-plugins**
      🗋 **net.wequick.small.properties**

FIGURE 10.23   Structure of buildSrc.

At last, in *build.gradle* of the app, we apply a plug-in to *net.wequick. small*:

```
apply plugin: "net.wequick.small"
```

Click the button "Sync Project With Gradle Files" in Android Studio, a new task named *testPlugin* will occur in the Gradle panel. *testPlugin* is created in *MyPlugin*; the definition of *MyPlugin* is as follows.

```
public class MyPlugin implements Plugin<Project> {
  @Override
  void apply(Project project) {
    project.extensions.create("pluginSrc", MyExtension)

    project.task("testPlugin") << {
      println project.pluginSrc.message
    }
  }
}
```

Click *testPlugin* in the Gradle panel; it will print log in the Gradle Console Panel, shown as follows:

```
Incremental Java compilation is an incubating feature.
:testPlugin
Hello gradle plugin in src

BUILD SUCCESSFUL

Total time: 2.11 secs
```

Up until now, we have finished the first demo.

Don't add the following declaration in *build.gradle* of the root directory in TestSmallGradle1.0; otherwise, only two tasks *testPlugin* and *clean* are shown in the Gradle panel.

```
apply plugin: "net.wequick.small"
```

10.7.1.2 Extension*

In Section 10.7.1.1, we have defined a task named *testPlugin* in *MyPlugin*; it prints a message. It's a fixed value. But we expect that the task can print a dynamic value. We can use *Extension*.

First, we create *MyExtension* in the project buildSrc, shown as follows:

```
class MyExtension {
  String message
}
```

Next, we use *MyExtension* in *MyPlugin*; the code is as follows:

```
public class MyPlugin implements Plug-in<Project> {

  @Override
  void apply(Project project) {
    project.extensions.create("pluginSrc", MyExtension)

    project.task("testPlugin") << {
      println project.pluginSrc.message
    }
  }
}
```

In the code above, we create a new *Extension*, its name is *pluginSrc*. So, we can use the following code in the project:

```
project.pluginSrc.message.
```

Next, in *build.gradle* of the app, we apply this new plug-in *net.wequick. small* to the current project and set a value for *pluginSrc.message*.

---

* For demo codes, go to https://github.com/Baobaojianqiang/ TestSmallGradle1.1

```
apply plugin: "net.wequick.small"

pluginSrc {
  message = "hello gradle plugin"
}
```

Now we execute the task *testPlugin*; we can find the message "hello Gradle plug-in" is printed.

10.7.1.3 The Hook App Packaging Process*

Gradle has a lot of internal tasks, such as *preBuild*. In the packaging process of the app, Gradle will execute these tasks one by one.

Let's create an Android project and execute the task *assembleRelease* in the Gradle panel; an *apk* will be generated. We can see the following logs in the Gradle console panel (Figure 10.24).

assembleRelease is the simplest task in the app packaging process. I list some important tasks in the following table.

Task Name	Functionality
processReleaseResources	*Exec aapt* to generate a *zip* file, and *R.java*
compileReleaseJavaWithJavac	*Exec javac*. Compile Java code to multiple .*class* files

Now let's have a look at the method *afterEvaluate()*, shown as follows:

```
public class MyPlugin implements Plug-in<Project> {

  @Override
  void apply(Project project) {
    //omit some code
    project.afterEvaluate() {
      def preBuild = project.tasks['preBuild']
      preBuild.doFirst {
        println 'hookPreReleaseBuild'
      }
      preBuild.doLast {
        println 'hookPreReleaseBuild2'
      }
    }
  }
}
```

---

* For demo codes, go to https://github.com/Baobaojianqiang/ TestSmallGradle1.2

```
:app:preBuild UP-TO-DATE
:app:preReleaseBuild UP-TO-DATE
:app:checkReleaseManifest
:app:preDebugBuild UP-TO-DATE
:app:prepareComAndroidSupportAnimatedVectorDrawable2520Library
:app:prepareComAndroidSupportAppcompatV72520Library
:app:prepareComAndroidSupportSupportCompat2520Library
:app:prepareComAndroidSupportSupportCoreUi2520Library
:app:prepareComAndroidSupportSupportCoreUtils2520Library
:app:prepareComAndroidSupportSupportFragment2520Library
:app:prepareComAndroidSupportSupportMediaCompat2520Library
:app:prepareComAndroidSupportSupportV42520Library
:app:prepareComAndroidSupportSupportVectorDrawable2520Library
:app:prepareReleaseDependencies
:app:compileReleaseAidl
:app:compileReleaseRenderscript
:app:generateReleaseBuildConfig
:app:generateReleaseResValues
:app:generateReleaseResources
:app:mergeReleaseResources
:app:processReleaseManifest
:app:processReleaseResources
:app:generateReleaseSources
:app:incrementalReleaseJavaCompilationSafeguard
:app:javaPreCompileRelease
:app:compileReleaseJavaWithJavac
:app:compileReleaseJavaWithJavac - is not incremental (e.g. outputs have
 changed, no previous execution, etc.).
:app:compileReleaseNdk UP-TO-DATE
:app:compileReleaseSources
:app:lintVitalRelease
:app:mergeReleaseShaders
:app:compileReleaseShaders
:app:generateReleaseAssets
:app:mergeReleaseAssets
:app:transformClassesWithDexForRelease
:app:mergeReleaseJniLibFolders
:app:transformNativeLibsWithMergeJniLibsForRelease
:app:transformNativeLibsWithStripDebugSymbolForRelease
:app:processReleaseJavaRes UP-TO-DATE
:app:transformResourcesWithMergeJavaResForRelease
:app:packageRelease
:app:assembleRelease
```

FIGURE 10.24 Console log when executing *assembleRelease*.

The app packaging process is to execute a series of internal tasks such as *preBuild*. We can get the task such as *preBuild* in the method *afterEvaluate()*.

After we get this task, we can invoke the method *doFirst()* or *doLast()* to execute some logic before or after this task is executed.

Now I have introduced the basic knowledge of a gradle plug-in. In Section 10.7.2, I will talk about how to hook the app packaging process.

10.7.2 Modify *resources.arsc*

In Section 8.3.1, I introduced a scenario. When we merge all the resource IDs of the HostApp and the plug-in together, the resource ID may conflict between these two apks. The solution is to rewrite the command *aapt* to specify different *PackageIds* for different plug-ins.

In this section, I will introduce another solution. After the command *aapt* is executed, we modify *R.java* and *resources.arsc* in the Gradle plug-in.

10.7.2.1 How to Find Resources in Android

Let's have a look at how to search for resources in the Android system. There are two classes involved in this process, as follows:

1) *Resources* find the file name of the resource by resource ID. It is a 1:1 mapping between the resource ID and the file name. This mapping is stored in *resources.arsc. Resources* search this file to find the corresponding file name.

2) *AssetManager* finds the resource by the resource name.

Figure 10.25 shows this process in detail.

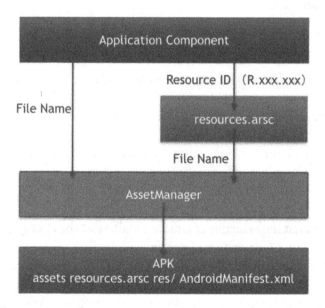

FIGURE 10.25 How to get resources in *AssetManager* and *Resources*.

10.7.2.2 Function of aapt

What does the command *aapt* do in the app packaging process? Refer to Figure 10.26 for details.

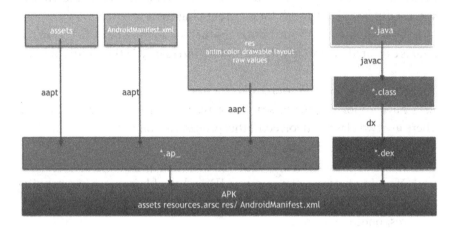

FIGURE 10.26   Generation after executing *aapt*.

In Figure 10.26, there are three steps, shown as follows:

1) Zip all the resources into a file.

2) Generate the resource ID for each resource and save this 1:1 mapping between the resource ID and the file name in *resources.arsc.*

3) Write resource ID for each resource in *R.java.*

After the command *aapt* is executed, it will execute *javac*, which will compile all Java files include *R.java.*

We can add our own logic after the command *aapt* has been executed and before the command *javac* has been executed.

10.7.2.3 The Principle of Gradle-Small

Let's talk about the principle of Gradle-Small: reset the *PackageId* for each resource ID in *R.java* and *resources.arsc.*

We also find there are a lot of redundant resources in *resources.arsc.* The packages like *AppCompat* and *Design* are stored in *resources.arsc.* It means if there are two plug-ins in the HostApp, the packages AppCompat and Design will have three duplicate copies.

The function of Gradle-Small is to remove these redundant resources in the plug-in; we keep the packages like *AppCompat* and Design only in the HostApp.

10.7.2.4 How to Use Gradle-Small

Because Gradle-Small has been uploaded to *JCenter*, we can use it by declaring it in *build.gradle*. On the other hand, Gradle-Small is open source, so we can include it in our plug-in project and use it directly after we compile the current project.

Gradle-Small supplies some tasks for us, and we can find them in the Gradle panel of Android Studio, shown as follows:

- *buildBundle*: It's used in plug-in packaging.

- *buildLib*: It's used in class library packaging.

- *small*: It's also a task used to summarize the information of all the plug-ins, as shown in Figure 10.27, the *PackageId* of *lib* is 0x79.

```
| type |    name    |  PP  | sdk |  aapt  | support | file(armeabi) | size |
|------|------------|------|-----|--------|---------|---------------|------|
| host | app        |      | 25  | 25.0.2 | 25.3.1  |               |      |
| app  | app.about  | 0x5a | 25  | 25.0.3 | 25.3.1  |               |      |
| lib  | lib.style  | 0x79 | 25  | 25.0.2 | 25.3.1  |               |      |
```

FIGURE 10.27   Outputs after executing the task small.

10.7.2.5 The Family of Plug-Ins Defined in Gradle-Small

There is a lot of custom plug-ins and *Extensions* in Gradle-Small, as shown in Figures 10.28 and 10.29.

Let's focus on one family of plug-ins, which is the core of Gradle-Small.

1) *BasePlugin*

BasePlugin is an ancestor of the plug-in family. It divides the method *apply()* into three methods, shown as follows:

- *createExtension()*
- *configureProject()*
- *createTask()*

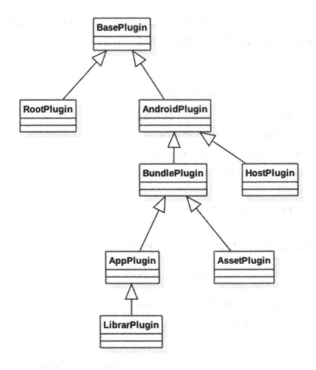

FIGURE 10.28   Family of plug-ins.

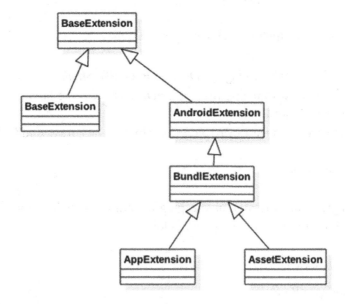

FIGURE 10.29   Family of *Extension*.

```
public abstract class BasePlugin implements
Plugin<Project> {
  void apply(Project project) {
    this.project = project

    createExtension()
    configureProject()
    createTask()
  }
}
```

The descendants of *BasePlugin* will override these three methods.

2) RootPlugin

RootPlugin is the entry point of Gradle-Small; it takes the role of function *main()*. In the method *configureProject()* of *RootPlugin*, it dispatches the request to the other plug-ins, shown as follows:

- *HostPlugin*: Correspond to the project app; in this scenario, its HostApp.

- *AppPlugin*: Correspond to the project which has a prefix app, such as *app.about*, in this scenario, all the plug-ins belong to *AppPlugin*.

- *LibraryPlugin*: Corresponds to the project which has a prefix *lib*, such as *lib.style*, in this scenario, it's a class library used by the HostApp and plug-in app.

- *AssetPlugin*: Corresponds to the project which only has resources. This project does not have any code.

I list four plug-in classes corresponding to four files in the folder *META-INF* in Figure 10.30.

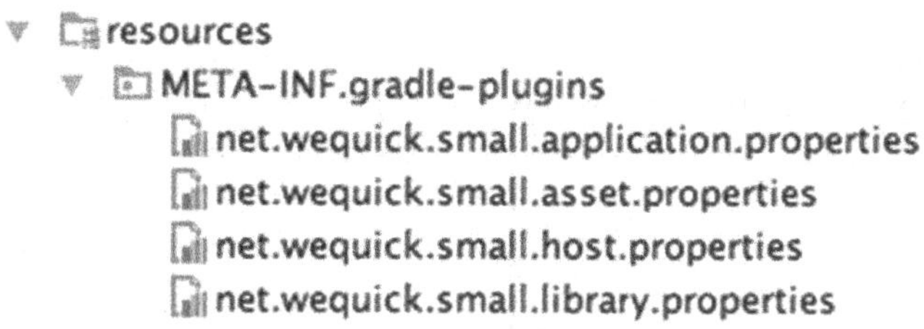

```
▼ ⌂ resources
    ▼ ⌂ META-INF.gradle-plugins
            ⌂ net.wequick.small.application.properties
            ⌂ net.wequick.small.asset.properties
            ⌂ net.wequick.small.host.properties
            ⌂ net.wequick.small.library.properties
```

FIGURE 10.30 Files in *META-INF*.

In the method *configureProject()* of *RootPlugin*, it starts the compiling process of *AppPlugin* or *LibraryPlugin*; the code is shown as follows:

```
switch (type) {
  case "App":
    it.apply plugin: AppPlugin
    rootExt.appProjects.add(it)
    break;
  case "lib":
    it.apply plugin: LibraryPlugin
    rootExt.libProjects.add(it)
    break;
```

3) *LibraryPlugin*

LibraryPlugin is used to handle the class library in the Android app.

4) *AppPlugin*

This is the core class for resolve the conflict of resource ID.

The method *initPackageId()* of *AppPlugin* will generate a new *PackageId* for each plug-in, such as 0x79. Of course, we can configure this *PackageId* manually in *Extension*. The value will be saved in a global array named *sPackageIds*; we will use it later.

The method *hookVariantTask()* of *AppPlugin* is responsible for hooking the Android app packaging process, shown as follows:

```
protected void hookVariantTask(BaseVariant variant) {
  hookMergeAssets(variant.mergeAssets)
  hookProcessManifest(small.processManifest)
  hookAapt(small.aapt)
  hookJavac(small.javac, variant.buildType.
  minifyEnabled)

  def mergeJniLibsTask = project.tasks.
  withType(TransformTask.class).find {
    it.transform.name == 'mergeJniLibs' &&
    it.variantName == variant.name
  }
```

```
hookMergeJniLibs(mergeJniLibsTask)

// Hook clean task to unset package id
project.clean.doLast {
  sPackageIds.remove(project.name)
  }
}
```

The method *hookVariantTask()* executes five methods which have the prefix *hook*. Let's introduce the most important methods as follows:

- *hookAapt()*

This method intercepted the original task *processReleaseResources*. It means we can add our own logic after the command *aapt* is executed.

aapt generates *R.java*; this file defines the resource ID for each resource; we change the *PackageId* of each resource to 0x79. And then we put *R.java* in a new location; we will use this file in the method *hookJavac()*. I will introduce this later.

aapt also produces a package, including *AndroidManifest.xml*, *resources. arsc*, and all the resources. We unzip this package and fetch *resources.arsc* from it. We change the *PackageId* of each resource in *resources.arsc* to 0x79.

This method also removes libraries like *AppCompat* and *Design*.

- *hookJavac()*

This method intercepted the original task *compileReleaseJavaWith-Javac*. It means we can remove some resource files such as *R$drawable. class* after the Java code is compiled to the *.class* files by the command *javac*. We must delete these files because the resource IDs stored in these files have the same prefix 0x7f.

In the method *hookAapt()*, we change the *PackageId* to 0x79 in *R.Java* and put this file in a new location. After we delete the files like *R$drawable. class*, we can execute the command *javac* to compile *R.Java* to *R.class*.

10.7.2.6 The Family of Editors Defined in Gradle-Small
Finally, let's have a look at how to modify *resources.arsc*. It's implemented by a series of Editors in Gradle-Small, as shown in Figure 10.31.

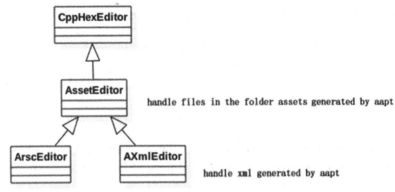

handle files in the folder assets generated by aapt

handle xml generated by aapt

handle resources. arsc generated by aapt

FIGURE 10.31   Family of Editors.

ArscEditor is the most important Editor in Figure 10.31.

Resource IDs are saved in the sections "Package Header" and "Type Spec & Type Info" in Figure 10.32. We change the *PackageId* from 0x7f to 0x79.

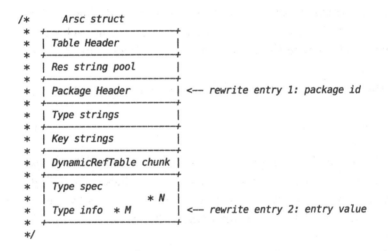

```
/*        Arsc struct
 *   +------------------------+
 *   | Table Header           |
 *   +------------------------+
 *   | Res string pool        |
 *   +------------------------+
 *   | Package Header         |  <-- rewrite entry 1: package id
 *   +------------------------+
 *   | Type strings           |
 *   +------------------------+
 *   | Key strings            |
 *   +------------------------+
 *   | DynamicRefTable chunk  |
 *   +------------------------+
 *   | Type spec              |
 *   |              * N        |
 *   | Type info  * M         |  <-- rewrite entry 2: entry value
 *   +------------------------+
 */
```

FIGURE 10.32   Structure of *resources.arsc*.

Refer to the author's blog for detail*. We won't spend much time on it.

It's difficult to read the source code of Gradle-Small. My suggestion is to write logs at the beginning and end of each method of Gradle-Small.

---

* Author's blog: https://github.com/wequick/Small/tree/master/Android/DevSample/buildSrc

When we execute a task generated by Gradle-Small, we can find how many methods are invoked from logs, shown as follows:

```
private def hookAapt(ProcessAndroidResources aaptTask) {
  Console.println('AppPlugin_hookAapt')
  Console.println(aaptTask)

  aaptTask.doLast { ProcessAndroidResources it ->
    // Unpack resources.ap_
    File apFile = it.packageOutputFile
```

10.8 COMPATIBILITY WITH ANDROID O AND P

In the previous chapters, I introduced a plug-in technique based on Android 7.0. All the samples can be found on my GitHub. When you run the demo on your phone with the Android O or P system, you may get an exception during runtime. This section discusses how to fix these bugs in these two Android systems.

Note: I list some popular mappings between Android system names, versions, and API level in the following table:

Android System	Version	API level
P		28
Oreo	8.1	27
Oreo	8.0	26
Nougat	7.1	25
Nougat	7.0	24
Marshmallow	6.0	23

Note: for the sample code refer to https://github.com/BaoBaoJianqiang/ZeusStudy1.2.

Switch to branch api26+ and have a look at the log as follows:

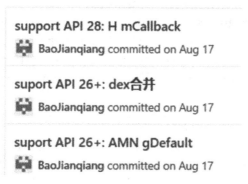

support API 28: H mCallback

BaoJianqiang committed on Aug 17

suport API 26+: dex合并

BaoJianqiang committed on Aug 17

suport API 26+: AMN gDefault

BaoJianqiang committed on Aug 17

10.8.1 Compatibility with Android O

In this section, we focus on the Android O system. We will discuss the modifications from Android N to O and introduce a solution for the compatibility of the Android O system.

10.8.1.1 Refactor of AMN

ActivityManagerNative (AMN) is modified in every Android system.

AMN has a field *gDefault*; its definition in API 25 is as follows:

```
public abstract class ActivityManagerNative extends
Binder implements IActivityManager {
  private static final Singleton<IActivityManager>
  gDefault = new Singleton<IActivityManager>() {
    protected IActivityManager create() {
      IBinder b = ServiceManager.getService("activity");
      if (false) {
        Log.v("ActivityManager", "default service
        binder = " + b);
      }
      IActivityManager am = asInterface(b);
      if (false) {
        Log.v("ActivityManager", "default service = " +
        am);
      }
      return am;
    }
  };
}
```

We can obtain the field *gDefault* of *AMN* by reflection, execute the method *create()* of *gDefault*. The method *create()* returns an object which implements the interface *IActivityManager*.

We can hook this *IActivityManager* object by the method *Proxy.new-ProxyInstance()*, intercept its method *startActivity()*, and replace the original *Activity* which is not declared in *AndroidManifest.xml* with a *StubActivity* declared in *AndroidManifest.xml*. The code is as follows:

```
public static void hookAMN() throws
ClassNotFoundException,
    NoSuchMethodException, InvocationTargetException,
    IllegalAccessException, NoSuchFieldException {
```

```
//Gets the gDefault singleton of AMN, which is
final and static
Object gDefault = RefInvoke.
getStaticFieldObject("android.app.
ActivityManagerNative", "gDefault");
// The gDefault is an android.util.singleton <T>
object; We pull out the mInstance field in this
singleton
Object mInstance = RefInvoke.
getFieldObject("android.util.Singleton", gDefault,
"mInstance");

// Create a proxy object for this object,
MockClass1, and replace the field with our proxy
object to help with the work
Class<?> classB2Interface = Class.
forName("android.app.IActivityManager");
Object proxy = Proxy.newProxyInstance(
    Thread.currentThread().getContextClassLoader(),
    new Class<?>[] { classB2Interface },
    new MockClass1(mInstance));

//Change gDefault's mInstance field to proxy
Class class1 - gDefault.getClass();
RefInvoke.setFieldObject("android.util.Singleton",
gDefault, "mInstance", proxy);
}
```

We introduced the logic in the above code in Chapter 5 in detail. Unfortunately, it doesn't work on Android O (API level 26), and the value of *gDefault* is *null*, shown as follows:

```
Object gDefault = RefInvoke.getStaticFieldObject
("android.app.ActivityManagerNative", "gDefault");
```

In Android O, the field *gDefault* is removed from *AMN*. Another field *IActivityManagerSingleton* defined in *ActivityManager* plays the same role as *gDefault*. This means we can use this field instead of *gDefault*, shown as follows:

```
Object gDefault = RefInvoke.
getStaticFieldObject("android.app.ActivityManager",
"IActivityManagerSingleton");
```

To be compatible with all Android systems, we use an "*if...else...*" statement to handle the differences among different Android scenarios, shown as follows:

```
Object gDefault = null;
    if (android.os.Build.VERSION.SDK_INT <= 25) {
        //Gets the gDefault singleton of AMN, which
        is static
        gDefault = RefInvoke.getStaticFieldObject("android.
        app.ActivityManagerNative", "gDefault");
    } else {
        //Obtain the ActivityManager singleton
        IActivityManagerSingleton, he really is the
        gDefault before
        gDefault = RefInvoke.getStaticFieldObject("android.
        app.ActivityManager", "IActivityManagerSingleton");
    }
```

10.8.1.2 The Story of Element *and* DexFile

Now let's review the process of how to load the class of a plug-in. In this book, I have introduced three mechanisms as follows:

1) Create a *ClassLoader* for each plug-in, and use the corresponding *ClassLoader* to load the classes of the plug-in.

2) Combine all the *dexes* of the plug-ins into the *dex* array of the HostApp.

3) The HostApp uses its original *ClassLoader* to load classes in the HostApp. The HostApp can't use its original *ClassLoader* to load classes in the plug-in app. So we write a new *ClassLoader* and replace the original *ClassLoader* with this new *ClassLoader*. This new *ClassLoader* has a collection storing all the *ClassLoaders* for each plug-in. It will try to load a class in the HostApp at first, and if not found it will iterate its internal collection to find which *ClassLoader* can load this class.

The second mechanism is the simplest solution, shown as follows:

```
public final class BaseDexClassLoaderHookHelper {

    public static void patchClassLoader(ClassLoader cl,
    File apkFile, File optDexFile)
```

```
  throws IllegalAccessException,
  NoSuchMethodException, IOException,
  InvocationTargetException,
  InstantiationException, NoSuchFieldException {
// Obtain BaseDexClassLoader : pathList
Object pathListObj = RefInvoke.
getFieldObject(DexClassLoader.class.
getSuperclass(), cl, "pathList");

// obtain PathList: Element[] dexElements
Object[] dexElements = (Object[]) RefInvoke.
getFieldObject(pathListObj, "dexElements");

// Element type
Class<?> elementClass = dexElements.getClass().
getComponentType();

// Create an array to replace the original array
Object[] newElements = (Object[]) Array.
newInstance(elementClass, dexElements.length + 1);

// Construct the plugin Element(File file, boolean
isDirectory, File zip, DexFile dexFile)
Class[] p1 = {File.class, boolean.class, File.
class, DexFile.class};
Object[] v1 = {apkFile, false, apkFile, DexFile.
loadDex(apkFile.getCanonicalPath(), optDexFile.
getAbsolutePath(), 0)};
Object o = RefInvoke.createObject(elementClass,
p1, v1);

Object[] toAddElementArray = new Object[] { o };
// The original elements are copied in
System.arraycopy(dexElements, 0, newElements, 0,
dexElements.length);
// Copy the element of the plugin
System.arraycopy(toAddElementArray, 0,
newElements, dexElements.length,
toAddElementArray.length);

// replace
```

```
    RefInvoke.setFieldObject(pathListObj,
    "dexElements", newElements);
  }
}
```

In the code above, let's pay attention to the following code lines, we execute the constructor of *Element*, but unfortunately, in the Android O system, this constructor is discarded:

```
Class[] p1 = {File.class, boolean.class, File.class,
DexFile.class};
Object[] v1 = {apkFile, false, apkFile, DexFile.
loadDex(apkFile.getCanonicalPath(), optDexFile.
getAbsolutePath(), 0)};
Object o = RefInvoke.createObject(elementClass, p1, v1);
Object[] toAddElementArray = new Object[] { o };
```

In addition, *DexFile*, which is used as a parameter in this constructor, is also discarded in Android O. Google explains that *DexFile* is only available inside the Android system, not open to app developers.

We can execute the method *makeDexElements()* of the *DexPathList* class to generate *dex* in the plug-in.

```
  List<File> legalFiles = new ArrayList<>();
  legalFiles.add(apkFile);

List<IOException> suppressedExceptions = new
ArrayList<IOException>();

  Class[] p1 = {List.class, File.class, List.class,
  ClassLoader.class};
  Object[] v1 = {legalFiles, optDexFile,
  suppressedExceptions, cl};
  Object[] toAddElementArray = (Object[])
  RefInvoke.invokeStaticMethod("dalvik.system.
  DexPathList", "makeDexElements", p1, v1);
```

The code above is also available in the previous versions of the Android system. This means we find a better method suitable for all the Android systems.

10.8.2 Compatibility with Android P

The impact of Android P on plug-ins is mainly in two aspects. One is the modification of the class *H*, and the other is the modification of the class *Instrumentation*.

10.8.2.1 The Modification of the Class H

10.8.2.1.1 Starting from *Message* and *Handler*

Android developers are familiar with two classes: *Message* and *Handler*. Let's have a quick review of how to send or receive a message in the Android system.

When the Android system starts a new app process, it will create *ActivityThread* at first; *ActivityThread* is the main thread and is also called the UI thread. The function *main()* of the app doesn't exist in the app itself; it's "hidden" in *ActivityThread*,

In the function *main()*, *MainLooper* is created. *MainLooper* is an endless loop that is responsible for receiving messages.

The class *Message* is defined as follows. It has three fields, *what* and *obj* is open to app developers, but *target* is a private field, and it's a *Handler* object.

```
public final class Message implements Parcelable {
  public int what;
  public Object obj;
  Handler target;

  //ignore some code
}
```

In the app process, each lifecycle method of the *Application* and the four components communicates with the *AMS* process frequently, as follows:

- The app process passes the data to the *AMS* process by *AMN*.

- The *AMS* process passes the data to the app process and the app uses *ApplicationThread* to receive the data from the *AMS*.

ApplicationThread invokes the method *sendMessage()* after it receives data. The method *sendMessage()* will send a message to the endless loop of *MainLooper*.

Now let's analyze how to handle the message in the *MainLooper*. Each message has a field *target*. When a message is thrown into the *MainLooper*, the Android system will invoke the method *dispatchMessage()* of the *target* to dispatch the message to the corresponding class to handle it. The field *target* is a *Handler* object.

It's time to have a look at the structure of the class *Handler*, shown as follows:

```
public class Handler {
  final Callback mCallback;

  public interface Callback {
    public boolean handleMessage(Message msg);
  }

  public void handleMessage(Message msg) {
  }

  public void dispatchMessage(Message msg) {
    if (msg.callback != null) {
      handleCallback(msg);
    } else {
      if (mCallback != null) {
        if (mCallback.handleMessage(msg)) {
          return;
        }
      }
      handleMessage(msg);
    }
  }
}
```

The method *handleMessage(Message msg)* of *Handler* has no logic, so we always write a subclass which inherits from *Handler* and overrides the method *handleMessage()*.

In the Android system, this subclass is *H*. *H* is an inner class defined in *ActivityThread*, shown as follows:

```
public final class ActivityThread {
  private class H extends Handler {
    public static final int LAUNCH_ACTIVITY    = 100;
```

```
public static final int PAUSE_ACTIVITY        = 101;
public static final int PAUSE_ACTIVITY_FINISHING=
102;
public static final int STOP_ACTIVITY_SHOW    = 103;
public static final int STOP_ACTIVITY_HIDE    = 104;
public static final int SHOW_WINDOW        = 105;
public static final int HIDE_WINDOW        = 106;
public static final int RESUME_ACTIVITY       = 107;
public static final int SEND_RESULT        = 108;
public static final int DESTROY_ACTIVITY      = 109;
public static final int BIND_APPLICATION      = 110;
public static final int EXIT_APPLICATION      = 111;
public static final int NEW_INTENT         = 112;
public static final int RECEIVER           = 113;

//ignore some code

public void handleMessage(Message msg) {
  switch (msg.what) {
    case LAUNCH_ACTIVITY: {
      final ActivityClientRecord r =
      (ActivityClientRecord) msg.obj;

      r.packageInfo = getPackageInfoNoCheck(
          r.activityInfo.applicationInfo,
          r.compatInfo);
      handleLaunchActivity(r, null,
      "LAUNCH_ACTIVITY");
    } break;

  //ignore some code
  }
 }
  }
}
```

The method *handleMessage(Message msg)* of *H* is used to dispatch the message to the different branches; for example, in the above code snippets, if *msg.what* is equal to *LAUNCH_ACTIVITY*, it will launch an *Activity*.

LAUNCH_ACTIVITY is a const, equal to 100. There are a lot of consts declared in *ActivityThread*. For example, *PAUSE_ACTIVITY* is equal to 101. From 100 to 109, all these consts are prepared for the lifecycle method of *Activity*. From 110, they are prepared for *Application, Service, ContentProvider*, and *BroadcastReceiver*.

Up until now, we have discussed the process of sending and receiving messages in Android. We need to remember the following two points:

- There is a field *mCallback* in the class *Handler*.

- *H* is used to handle different messages.

10.8.2.1.2 Plug-In solutions before Android P

Before Android P (API level 28), we hook the field *Callback* of *H* and intercepted the method *handleMessage()* of *H*, shown as follows:

```
class MockClass2 implements Handler.Callback {

  Handler mBase;

  public MockClass2(Handler base) {
    mBase = base;
  }

  @Override
  public boolean handleMessage(Message msg) {

    switch (msg.what) {
      case 100:
        handleLaunchActivity(msg);
        break;
    }

    mBase.handleMessage(msg);
    return true;
  }

  private void handleLaunchActivity(Message msg) {
    Object obj = msg.obj;

    Intent raw = (Intent) RefInvoke.
    getFieldObject(obj, "intent");

    Intent target = raw.getParcelableExtra(AMSHookHelp
    er.EXTRA_TARGET_INTENT);
```

```
    raw.setComponent(target.getComponent());
  }
}
```

We have discussed *MockClass2* in Section 5.2.4.

10.8.2.1.3 New message mechanism in Android P

Android P refactors the process of messages used in *Activity*. We have introduced that there are ten messages defined in *ActivityThread* from 100 to 109 and prepared for *Activity*. These ten messages are missing in *ActivityThread*; there is a new message *EXECUTE_TRANSACTION* equal to 159 defined in *ActivityThread*.

Why did Google make this change? Because *Activity* has a lot of lifecycle methods, and these methods form a state machine, shown in Figure 10.33.

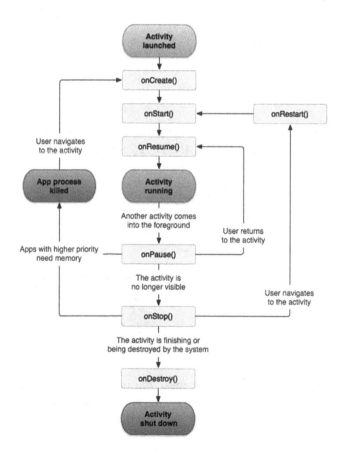

FIGURE 10.33   State machine of lifecycle methods in *Activity*.

It's not convenient to write a lot of branches in the method *handleMessage()* of *H*, shown as follows:

```
private class H extends Handler {
  public void handleMessage(Message msg) {
    switch (msg.what) {
      case LAUNCH_ACTIVITY: {
        final ActivityClientRecord r =
        (ActivityClientRecord) msg.obj;

        r.packageInfo = getPackageInfoNoCheck(
          r.activityInfo.applicationInfo,
          r.compatInfo);
        handleLaunchActivity(r, null,
        "LAUNCH_ACTIVITY");
      } break;

    //ignore some code
      }
    }
}
```

We create a class for each state; for example, we create *LaunchActivityItem* to replace the branch *LAUNCH_ACTIVITY*, shown in Figure 10.34.

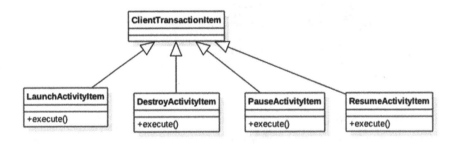

FIGURE 10.34 *Activity* lifecycle family.

In Android P, we can remove these ten branches in the method *handleMessage()* of *H*, and the implementation of *LaunchActivityItem* is as follows:

```
public class LaunchActivityItem extends
ClientTransactionItem {

  @Override
  public void execute(ClientTransactionHandler client,
  IBinder token,
     PendingTransactionActions pendingActions) {
    ActivityClientRecord r = new
    ActivityClientRecord(token, mIntent, mIdent,
    mInfo,
      mOverrideConfig, mCompatInfo, mReferrer,
      mVoiceInteractor, mState, mPersistentState,
        mPendingResults, mPendingNewIntents,
        mIsForward,
        mProfilerInfo, client);
    client.handleLaunchActivity(r, pendingActions,
    null /* customIntent */);
  }
}
```

Although we create a lot of classes for each state, the code in the method *handleMessage()* of *H* is less. It fits in the open–close principle. We can have more than 100 classes, but we can't write all the code into one class.

10.8.2.1.4 Hook class *H* in Android P

Android P deletes ten messages from 100 to 109 in the method *handleMessage()* of *H*, and adds a new message equal to 159, named *EXECUTE_TRANSACTION*.

This modification causes the plug-in framework not to work in Android P because we use *MockClass2* to intercept the message equal to 100, shown as follows:

```
class MockClass2 implements Handler.Callback {

  Handler mBase;

  public MockClass2(Handler base) {
    mBase = base;
  }

  @Override
  public boolean handleMessage(Message msg) {
```

```
  switch (msg.what) {
    case 100:
      handleLaunchActivity(msg);
      break;
  }

  mBase.handleMessage(msg);
  return true;
}
```

Now we need intercept the message equal to 159, and dispatch a different state to a different branch, shown as follows:

```
class MockClass2 implements Handler.Callback {

  Handler mBase;

  public MockClass2(Handler base) {
    mBase = base;
  }

  @Override
  public boolean handleMessage(Message msg) {

    switch (msg.what) {
      case 100:  //below API 28
        handleLaunchActivity(msg);
        break;
      case 159:  //for API 28
        handleActivity(msg);
        break;
    }

    mBase.handleMessage(msg);
    return true;
  }

  private void handleActivity(Message msg) {
    Object obj = msg.obj;

    List<Object> mActivityCallbacks = (List<Object>)
    RefInvoke.getFieldObject(obj, "mActivityCallbacks");
```

```
if(mActivityCallbacks.size() > 0) {
  String className = "android.app.
  servertransaction.LaunchActivityItem";
  if(mActivityCallbacks.get(0).getClass().
  getCanonicalName().equals(className)) {
    Object object = mActivityCallbacks.get(0);
    Intent intent = (Intent) RefInvoke.
    getFieldObject(object, "mIntent");
    Intent target = intent.getParcelableExtra(AMSHo
    okHelper.EXTRA_TARGET_INTENT);
    intent.setComponent(target.getComponent());
  }.
}
}
}
```

10.8.2.2 *The Refactoring of the Class* Instrumentation

Before Android P, the logic of the method *newActivity()* of *Instrumentation* is as follows:

```
public Activity newActivity(ClassLoader cl, String
className, Intent intent)
    throws InstantiationException,
    IllegalAccessException, ClassNotFoundException {
  return (Activity)cl.loadClass(className).
  newInstance();
}
```

Android P rewrites the logic of *Instrumentation*. In the method *newActivity()* of *Instrumentation*, it will check the field *mThread* of *Instrumentation*. If it is empty, it will throw an exception:

```
public class Instrumentation {
  public Activity newActivity(ClassLoader cl, String
  className,
    Intent intent)
    throws InstantiationException,
    IllegalAccessException,
    ClassNotFoundException {
  String pkg = intent != null && intent.
  getComponent() != null
```

```
      ? intent.getComponent().getPackageName() :
      null;
  return getFactory(pkg).instantiateActivity(cl,
  className, intent);
}

private AppComponentFactory getFactory(String pkg) {
  if (pkg == null) {
    Log.e(TAG, "No pkg specified, disabling
    AppComponentFactory");
    return AppComponentFactory.DEFAULT;
  }
  if (mThread == null) {
    Log.e(TAG, "Uninitialized ActivityThread, likely
    app-created Instrumentation,"
        + " disabling AppComponentFactory", new
        Throwable());
    return AppComponentFactory.DEFAULT;
  }
  LoadedApk apk = mThread.peekPackageInfo(pkg, true);
  // This is in the case of starting up "android".
  if (apk == null) apk = mThread.getSystemContext().
  mPackageInfo;
  return apk.getAppFactory();
  }
}
```

In Chapter 5, we introduced a hooking solution to intercept the method *execStartActivity()* of *Instrumentation*, as follows:

```
public class HookHelper {

  public static void attachContext() throws Exception{
    Object currentActivityThread = RefInvoke.
    invokeStaticMethod("android.app.ActivityThread",
    "currentActivityThread");

    Instrumentation mInstrumentation =
    (Instrumentation) RefInvoke.getFieldObject(current
    ActivityThread, "mInstrumentation");

    Instrumentation evilInstrumentation = new EvilInst
    rumentation(mInstrumentation);
```

```
  RefInvoke.setFieldObject(currentActivityThread,
    "mInstrumentation", evilInstrumentation);
  }
}

public class EvilInstrumentation extends
Instrumentation {

  private static final String TAG =
"EvilInstrumentation";

  Instrumentation mBase;

  public EvilInstrumentation(Instrumentation base) {
    mBase = base;
  }

  public ActivityResult execStartActivity(
      Context who, IBinder contextThread, IBinder
      token, Activity target,
      Intent intent, int requestCode, Bundle options) {

    Log.d(TAG, "hello guys");

    Class[] p1 = {Context.class, IBinder.class,
        IBinder.class, Activity.class,
        Intent.class, int.class, Bundle.class};
    Object[] v1 = {who, contextThread, token, target,
        intent, requestCode, options};

    return (ActivityResult) RefInvoke.
    invokeInstanceMethod(
        mBase, "execStartActivity", p1, v1);
  }
}
```

In the code above, we replace the original *Instrumentation* with *EvilInstrumentation*. But it doesn't work in Android P. It will throw an exception, as follows:

```
Uninitialized ActivityThread, likely app-created
Instrumentation.
```

We find the root cause of this exception, the field *mThread* of *EvilInstrumentation*, is empty.

To solve this bug, we must rewrite the method *newActivity()* of *EvilInstrumentation*, shown as follows:

```
public class EvilInstrumentation extends
Instrumentation {
  //ignore some code

  public Activity newActivity(ClassLoader cl, String
  className,
          Intent intent)
    throws InstantiationException,
    IllegalAccessException,
    ClassNotFoundException {

    return mBase.newActivity(cl, className, intent);
  }
}
```

When *EvilInstrumentation* executes its method *newActivity()*, the original method *newActivity()* of *Instrumentation* will be executed. The field *mThread* of *Instrumentation* is not *null*, so the exception above won't be thrown up.

Note: For the sample code refer to https://github.com/BaoBaoJianqiang/Hook15. Switch to the branch named *api26+*; the log is as follows:

Commits on Aug 19, 2018

support API 28: Hook execStartActivity

BaoJianqiang committed on Aug 19

In this chapter, I introduced the compatibility of plug-ins with Android O and P. As Google releases a new version of the Android system every year, we need to keep an eye on the changes which cause plug-in framework to stop functioning correctly.

10.9 SUMMARY

In this chapter, we have introduced a lot of problems with plug-in techniques. It's not only useful for plug-ins but widely used in other domains.

CHAPTER 11

Summary of Plug-In Technology

I N THE PREVIOUS CHAPTERS, we introduced the history of plug-in technology, different plug-in solutions, and a lot of well-known plug-in frameworks.

We'll review all these concepts and techniques in this summary chapter.

11.1 PLUG-IN ENGINEERING

There are always more than three projects in the plug-in: HostApp, MyPluginLibrary, Plugin1, and perhaps other plug-in projects.

The HostApp and the plug-in app all depend on MyPluginLibrary.

Chapter 6 describes how to create one solution with all the three projects in Android Studio; we can debug from the HostApp to Plugin1 or MyPluginLibrary.

Refer to Section 10.4 for plug-in signing and obfuscation.

Refer to Section 10.5 for incremental updates using plug-ins.

11.2 CLASS LOADING IN THE PLUG-IN

It is impossible to use the *ClassLoader* of the HostApp to load the classes of Plugin1. We have three solutions to load classes from Plugin1:

   1) Use the *ClassLoader* of Plugin1 to load classes from Plugin1. Refer to Chapter 6.

2) The app stores its *dexes* in an array. We combine all the *dexes* of all the plug-ins into the *dex* array of the HostApp. This means the *ClassLoader* of the HostApp can load any class whether this class in the HostApp or the plug-in. Refer to Section 8.2.3.

3) Write a custom *ClassLoader* to replace the original *ClassLoader* of the HostApp. In this custom *ClassLoader*, declare a collection to store all the *ClassLoaders* of the HostApp and the plug-ins. If we use this custom *ClassLoader* to load one class, it will traverse each *ClassLoader* stored in the collection to check which *ClassLoader* can load that particular class. Refer to Section 8.2.6.

11.3 WHICH CLASS OR INTERFACE CAN BE HOOKED?

"Once a lie starts, it can never be stopped. You need to continuously patch it up."

We always use this phrase to describe the *DroidPlugin* framework. *DroidPlugin* can nearly hook all the classes of the Android system.

Refer to Chapter 5 for more information about hooking technology. It's easy to identify which class we can hook, shown as follows:

1) Classes which are open to the app developers

Most of the classes and methods in the Android system are not open to app developers. If we want to use them, we must use reflection. We can't hook these classes or methods.

We can only hook the classes which are open to app developers. For example, we can hook *Instrumentation* and *Callback*, but we can't hook *H*.

We can create a class named *EvilInstrumentation*, which inherits *Instrumentation*; we can use an instance of *EvilInstrumentation* to replace the original instance of *Instrumentation*.

2) Classes which implement the interface

If one class implements an interface, we can use *Proxy.newProxyInstance()* to intercept the methods of this class. For example, *ActivityManagerNative* implements the interface *IActivityManager*, so we can hook *ActivityManagerNative*.

Another interface introduced in this book is *IPackageManager*.

3) Collection

Most of the classes and methods are not open to app developers, but if there is a collection field in the class we can add an element to this collection using reflection.

A typical example is to create a *LoadedApk* object and put it into the cache "in advance," as shown in Section 8.2.2.

11.4 A PLUG-IN SOLUTION FOR *ACTIVITY*

Generally, plug-in solutions can be divided into *Dynamic-Proxy* and *Static-Proxy*.

1) *Dynamic-Proxy* is referred to in Section 8.2. Create an empty *StubActivity* in the HostApp as a placeholder. Start *ActivityA* of the plug-in but tell the *AMS* to start the *StubActivity*. After cheating the *AMS* successfully, change the *StubActivity* to *ActivityA*. We need to hook a lot of classes to finish this mechanism.

2) *Static-Proxy* is referred to in Chapter 9. Create a *ProxyActivity* in the HostApp that determines which *Activity* of plug-in to launch. The *Activities* of the plug-in don't have a lifecycle, so in the lifecycle method of *ProxyActivity*, we invoke the corresponding lifecycle method of the plug-in.

In addition, we need support the *LaunchMode* of the *Activity* of the plug-in. Refer to Section 8.2.5 and 9.1.7

To launch the *Activity* from the plug-in, we also need to load the resources of the *Activity* of the plug-in. Let's review this technique in Section 11.5.

11.5 A PLUG-IN SOLUTION FOR *RESOURCES*

Resources are only used in an *Activity*.

The app loads the resources with the method *addAssetPath(String dexPath)* of *AssetManager*; the parameter *dexPath* is the location of the *apk*.

In the HostApp, the original *AssetManager* can only load the resources of the HostApp, and can't load the resources of Plugin1. We have two solutions for this limitation:

1) When we navigate from the HostApp to Plugin1, we create a new *AssetManager* to load the resources of Plugin1.

When we jump from Plugin1 to the HostApp, we switch to the original *AssetManager* to load resources from the HostApp.

Refer to Chapter 7 for the implementation of this solution.

2) In the HostApp, invoke the method *addAssetPath(String dexPath)* of *AssetManager*. If there are three plug-ins, we need to invoke this method three times. We set the parameter *dexPath* to the path of Plugin1, Plugin2, and Plugin3. After we invoke this method three times, the resources of the HostApp will contain all the resources, including from the HostApp, Plugin1, Plugin2, and Plugin3.

We need to store all these resources in a global variable so that the HostApp and plug-ins can visit them.

Refer to Section 9.1.4 for the implementation of this solution.

In solution 2, we need to merge the resource IDs of the HostApp and all the plug-ins. But the value of the resource ID of the HostApp may conflict with the value of the resource ID of the plug-in. We have two solutions to resolve this issue:

1) Modify the command *aapt* by adding a new argument for this command to specify the *PackageId* of the resource ID, such as 0x71. Refer to Section 10.1.2.

2) Hook the packaging process. After the command *aapt* is executed, the file *R.java* and *resources.arsc* are generated. We can modify these two files to replace the *PackageId* of the resource ID with 0x71. Refer to Section 10.7 for details.

11.6 USE *FRAGMENT* IN THE PLUG-IN

I Once confused the differences between *Fragment* and *Activity*.

The biggest difference is that *Activity* needs to communicate with the *AMS* frequently, but *Fragment* doesn't need to do that. The *Fragment* is just like a *CustomView* that lives in the *Activity*.

Because *Fragment* needn't communicate with the *AMS*, we needn't declare *Fragment* in the *AndroidManifest.xml*. Thus we have a new plug-in solution to write only one *Activity* in the app. All the pages are implemented by *Fragment*. This one *Activity* hosts all the *Fragments*, whether the *Fragment* comes from the HostApp or the plug-in.

For the plug-in solution based on *Fragment* refer to Section 10.2.

11.7 PLUG-IN SOLUTIONS FOR *SERVICE, CONTENTPROVIDER,* AND *BROADCASTRECEIVER*

The number of components in the app, including *Service*, *ContentProvider*, and *BroadcastReceiver*, is almost less than ten; we don't need to add a new component of a plug-in dynamically. So, we can declare these components in the *AndroidManifest.xml* in advance. Refer to Section 8.1.1.

The disadvantage of this solution is that we can't add a new component dynamically. However, we can handle these problems with the following solutions.

11.7.1 A Plug-In Solution for *Service*

There are three different plug-in solutions for *Service*:

1) *Dynamic-Proxy. Service* is different from *Activity*; we can declare one *StubActivity* in the *AndroidManifest.xml* of the HostApp to correspond to multiple *Activities* of the plug-ins. However, if we declare one *StubService* in the *AndroidManifest.xml* of the HostApp, it only corresponds to one *Service* of the plug-in. So, we need to pre-declare multiple *StubServices* to correspond to the multiple *Services* of the plug-ins. Refer to Section 8.1.3 for details.

2) *Static-Proxy.* Create *ProxyService* in the HostApp. We can use *ProxyService* to launch the *Service* of the plug-in. But one *ProxyService* can't correspond to multiple *Services*. So we should pre-declare multiple *ProxyServices*. Refer to Section 9.2.1, 14.2 and 14.3.

3) The last solution is to declare one *StubService* to correspond to multiple *Services* of the plug-in. Refer to Section 9.2.4 for details.

11.7.2 A Plug-In Solution for *BroadcastReceiver*

BroadcastReceiver has two types: *Dynamic Receiver* and *Static Receiver*.

Dynamic Receiver is simple. It's only a simple class that has one method *onReceive()*. We can use *ClassLoader* to easily load *Dynamic Receiver* by reflection.

Static Receiver is different from *Dynamic Receiver*. We must declare *Static Receiver* in the *AndroidManifest.xml*. We can send a broadcast to the *Static Receiver*; even if the app is not launched.

We have a lot of plug-in solutions to support *Static Receiver*.

1) The simplest plug-in solution for a *Static Receiver* is to transform the *Static Receiver* of the plug-in to a *Dynamic Receiver* registered in the HostApp. The disadvantage of this solution is that we can't use the features of the *Static Receiver*. If the app is not launched, we can send a broadcast to the *Static Receiver* of plug-in. Refer to Section 8.4.3 for details.

2) *Static-Proxy*. We declare a *ProxyReceiver* in the HostApp; it's responsible for dispatching the broadcast to the corresponding *Static Receiver* of the plug-in. Refer to Section 9.2.5 for details. This solution has the same disadvantage in that we can't use the features of the *Static Receiver*.

3) *Dynamic-Hook*. Declare one *StubReceiver* with multiple *Actions* in the *AndroidManifest.xml* of the HostApp. Each *Action* of the *StubReceiver* corresponds to a *Static Receiver* in the plug-in. In the method *onReceive()* of *StubReceiver*, it dispatches the broadcast to the corresponding *Static Receiver* of the plug-in. Refer to Section 8.4.4 for details.

11.7.3 A Plug-In Solution for *ContentProvider*

ContentProvider is the engine of SQLite. It provides CRUD methods for app developers.

In the plug-in solution for *ContentProvider*, we can pre-declare a *StubContentProvider* to cheat the *AMS*. *StubContentProvider* is open to the user. When the third-party user invokes the CRUD methods of the *StubContentProvider*, *StubContentProvider* will dispatch the request to the corresponding *ContentProvider* of the plug-in. Refer to Section 8.1.5 for details.

11.8 SUMMARY

In the last chapter of this book, we reviewed all the technologies involved in plug-ins, and we hope that readers can master this technique.

Appendix A: Sample Code List

THIS BOOK SUPPLIES OVER 70 code samples, and I list all the code samples here for convenience. The GitHub web address is long, so the same prefix is used for each project: https://github.com/baobaojianqiang/; you can view the sample at https://github.com/baobaojianqiang/<projectName>.

Chapter	Section	Project	Description
2	2.15.1	ReceiverTestBetween ActivityAndService1	Music player implemented by two receivers
	2.15.2	ReceiverTestBetween ActivityAndService2	Music player implemented by only one receiver
3	3.1	TestReflection	Reflection, basic syntax
	3.2	TestReflection2	Reflection using jOOR
	3.3	TestReflection3	Reflection encapsulated by the author of this book
	3.4	TestReflection4	Based on TestReflection3
4	4.2	InvocationHandler	*Dynamic-Proxy*
	4.3	HookAMS	Hook the *AMS*
	4.4	HookPMS	Hook the *PMS*
5	5.2.2	Hook11	Hook *mInstrumentation* of *Activity*
	5.2.3	Hook12	Hook *startActivity* of *AMN*
	5.2.4	Hook13	Hook the callback of *H*
	5.2.5	Hook14	Hook *Instrumentation*
	5.3.1	Hook15	Hook *Instrumentation*, based on *ActivityThread*
	5.4.2	Hook31	Start an *Activity* not declared in
	5.4.3		*theAndroidManifest.xml*, first-half
	5.4.4	Hook32	Start an *Activity* not declared in the *AndroidManifest.xml*, second-half

(Continued)

411

Chapter	Section	Project	Description
6	6.1	Dynamic0	Reflect a *dex* file
	6.2	Dynamic1.0	Interface-Oriented programming
	6.3	Dynamic1.1	Keyword "provided"
	6.4	Dynamic1.2	Debug
	6.5	ZeusStudy1.8	Plug-in for *Application*
7	7.2	Dynamic1.3	Read resource in the plug-in
		Dynamic2	Refactor based on Dynamic1.3
	7.3	Dynamic3	Skin Changing
	7.4	Dynamic3.2	Skin changing, another solution
8	8.1.1	ZeusStudy1.0	*Service* plug-in solution
	8.1.2		
	8.1.3		
	8.1.4	ZeusStudy1.1	*Activity* plug-in solution, not including *StubActivity*
	8.2.1	ActivityHook1	*Activity* plug-in solution 1
	8.2.2		
	8.2.3	ActivityHook2	*Activity* plug-in solution 2
	8.2.4	ZeusStudy1.2	Read *Resource* in the plug-in
	8.2.5	ZeusStudy1.3	*LaunchMode* solution
	8.2.6	ZeusStudy1.4	Hook *ClassLoader* with *ZeusClassLoader*
	8.3.2	ServiceHook1	*createService*
	8.3.3		
	8.3.4	ServiceHook2	*bindService*
	8.4.2	Receiver1.0	*Dynamic BroadcastReceiver* plug-in solution
	8.4.3	Receiver1.1	Static *BroadcastReceiver* plug-in solution
	8.4.4	Receiver1.2	Final plug-in solution for *Static BroadcastReceiver*
	8.5.2	ContentProvider1	*ContentProvider* demo
	8.5.3	ContentProvider2	*ContentProvider* plug-in solution
	8.5.4		
	8.5.5		
9	9.1.2	That1.0	The simple demo of the "That" framework
	9.1.3	That1.1	Navigation in plug-in inside
	9.1.4	That1.2	Keyword "that"
	9.1.5	That1.3	Navigation between the plug-in and HostApp

(Continued)

Chapter	Section	Project	Description
	9.1.6	That1.4	Interface-Oriented programming, using *IRemoteActivity*
	9.1.7	That1.5	Support *LaunchMode*
		TestSingleInstance	*singleInstance* demo
	9.2.1	That3.1	*startService*
	9.2.2	That3.2	*bindService*
	9.2.3	That3.3	*StubService*
	9.2.4	ServiceHook3	Another plug-in solution to *startService*
		ServiceHook4	Another plug-in solution to *bindService*
	9.2.5	That3.4	Use stub for *BroadcastReceiver*
10	10.1.2	AAPT	Modify *aapt*
		TestAAPTUpdate	Use new *aapt* command in App
	10.1.4	Apollo1.1	*public.xml*
	10.1.5	Apollo1.2	Plug-in using the resources in the HostApp
	10.2.2	Min18Fragment	Simple demo based on *Fragment* plug-in solution
	10.2.3	Min18Fragment2	Navigate from one plug-in to another plug-in
	10.2.4	Min18Fragment3	Navigation from the plug-in to the HostApp
	10.3	Hybrid1.2	Replace *Activity* with an *HTML5* page based on *startActivityForResult*
	10.4.1	Sign1	Basic Knowledge of ProGuard
		Sign2	Use keep in the HostApp
	10.4.2	ZeusStudy1.5	Not obfuscate *lib*
	10.4.3	ZeusStudy1.6	Obfuscate *lib*
	10.5	That2	Incremental update based on the plug-in
	10.6.1	JniHelloWorld	A simple SO file is written by the author of this book
	10.6.2	MySO1	Use SO from a third party, both 32 bits and 64 bits
	10.6.3	MySO1.1	Use SO from a third party, 32 bits only
		JniHelloWorld2	A simple SO file is written by the author of this book
		MySO2	Load the SO file in the HostApp. The SO files are located in the directory *assets*

(Continued)

414 ■ Appendix A

Chapter	Section	Project	Description
	10.6.4	MySO3	*system.loadLibrary*
	10.6.5	ZeusStudy1.7	*sysntem.load*
	10.7.1.1	TestSmallGradle1.0	Custom plug-in
	10.7.1.2	TestSmallGradle1.1	Custom Extension
	10.7.1.3	TestSmallGradle1.2	Hook the app building process in Gradle
	10.8.1.1	ZeusStudy1.2	*AMN*
	10.8.1.2		*Element* and *DexFile*
	10.8.2.1		*H* class
	10.8.2.2	Hook15	*execStartActivity* in *Instrumentation*

Index

23Code, 3
65536 issue, 64

aapt, 308, 385
  function, 380
  generation after executing, 380
  hooking
    modifying and generating new
      command, 308, 310–313
    new *appt* command in project,
      314–316
aapt_mac command, 314
aar files, 9
ABTest, 8
ACDD project, 5
Activity component, 10, 13, 23, 38, 65, 327;
    see also AMS
  code in, 72–74
  hooking of launching process of, 135
  plug-in solution for, 185–216
ActivityManagerNative (*AMN*), 29, 87, 406
  class diagram of, 29
  hooking of *getDefault()* method of,
    121–125
  proxy pattern and, 111–113
  refractor, 388–390
ActivityManagerNative.getDefault()
  method, 111
ActivityManagerNative.getDefault().start
  Activity() method, 29
ActivityManagerProxy (AMP) object, 29
  class diagram of, 29
ActivityManagerService, 88
ActivityNotFoundException, 30, 134
ActivityRecord object, 31

ActivityThread, 24, 27, 33, 34, 36, 43,
    50, 113, 285, 393, 394, 397;
    see also mInstrumentation
  communication with *H*, 125
  interaction with *H*, 36
  and *PackageManagerService* (PMS),
    60–61
  relationship with *H*, 126
addAssetPath() method, 1, 3, 6, 161,
  165, 319
addAssetPath(String dexPath)
  method, 408
addAssetPath(String path) method,
  161, 185
Add dependencies, between projects, 149
add method, 107–108
address field, 85
afterEvaluate() method, 377–378
AIDL, 13, 17–22, 308
  class diagram of, 18
  full picture of, 21
  with proxy pattern, 107–108
Ajax, 11
Alibaba, 7
Amazon app, 23, 24, 33, 34
  starting activity, 35–37
AMN.getDefault() method, 29
AMS, 14, 22–23, 126, 139, 393
  app sending messages to, 45
  hooking of, 133–135
  informing app, 46–47
  process from apps to, 117
  process from, to apps, 117
  Service component and, 42–47
  service sending broadcast to, 50

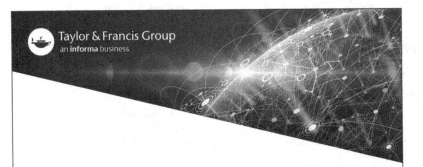